# HF Amateur Radio
## 2nd Edition

Ian Poole, G3YWX

Radio Society of Great Britain

Published by the Radio Society of Great Britain, Cranborne Road, Potters Bar, Herts EN6 3JE.

First Published 2001
Digitally Reprinted 2006
Second Edition 2007

© Radio Society of Great Britain, 2007. All rights reserved. No part of this publication may be reproduced, stored in a retrieval system, or transmitted, in any form or by any means, electronic, mechanical, photocopying, recording or otherwise, without the prior written permission of the Radio Society of Great Britain.

ISBN 9781-9050-8629-0

*Publisher's Note*

The opinions expressed in this book are those of the author and not necessarily those of the RSGB. While the information presented is believed to be correct, the author, the publisher and their agents cannot accept responsibility for consequences arising from any inaccuracies or omissions.

Cover Design: Dorotea Vizer, M3VZR
Production: Mark Allgar, M1MPA
DTP and Editing: Alex Kearns, M3LSZ

Printed in Great Britain by Latimer Trend of Plymouth.

Supporting website:
http://www.rsgb.org/books/extra/hfamradio.htm

# Table of Contents

|   | Preface | |
|---|---|---|
| 1 | An Introduction to the HF Bands | 1 |
| 2 | Radio Wave Propagation | 7 |
| 3 | Types of Transmission | 23 |
| 4 | Receivers | 35 |
| 5 | Transmitters | 55 |
| 6 | Antennas | 72 |
| 7 | Bands and Band Plans | 97 |
| 8 | On the Bands | 107 |
| 9 | Setting up the Radio Station | 121 |
|   | Appendix - Abbreviations and Codes | 133 |
|   | Index | 137 |

# Preface

THE HF or short wave bands are, to my mind, among the most fascinating areas of amateur radio. It is possible to make contacts with stations all over the world and chat to people you would not normally come across. In addition, there are many interesting challenges in setting up and operating a station to its maximum potential. Then, there's the excitement of contests, or contacting a station on a rare island or in a little heard of country. One never quite knows what the next station will be as you tune the dial. Coupled with this, the HF bands provide a wealth of opportunities to experiment with antennas, build your own equipment and tinker with new electronics and radio technologies.

Now with the relaxation of regulations, in particular the requirement for a Morse test, many more people have access to the HF bands. With far more people entering the fascinating world of HF amateur radio, I decided to update the HF Amateur Radio book I wrote some years ago. The aim of this new edition is to introduce newcomers to the hobby to the excitement of operating on the HF bands, explaining the techniques they need to master and knowledge they need to learn to get the most out of these bands,

I can still remember the thrill of making my first contact on 80m, then my first contact outside the UK, and later, early one morning, making contacts on 20m with the West Coast of the USA. Achieving contacts over these vast distances still gives me a sense of awe and excitement. Hopefully this book will help others, whether new to the hobby or somewhat more experienced, to gain more from amateur radio, and to experience as much pleasure from HF operating as I have over the years.

Despite all the new technology that is available today, HF band operating still has a lot to offer, not only in operating enjoyment and talking to people from all around the globe but also in exploring the limits of science and technology.

*Ian Poole*
*January 2007*

CHAPTER 1

# An Introduction to the HF Bands

*In this chapter:*
- Aspects of the hobby
- Amateur radio bands

THE short-wave (HF) bands provide one of the main areas of activity within amateur radio today. On these bands it is possible to hear and contact stations from all over the world, and even a newcomer can make QSOs with amateurs thousands of miles away.

Operating skills and know-how, combined with the right equipment, are all important to make the most of these fascinating bands. While a newcomer will be able to make interesting contacts, expert operators can get a lot more out of the bands, often contacting stations on unusual islands or on the other side of the globe. Their success can be attributed to experience. This may be in assembling, maintaining and improving the station but it could also be understanding when to listen on the bands or which frequencies to use at a particular time. It might also be knowing where to find the latest information about activity on the bands. Yet again, it may be knowing how to seek out rare stations and how to make contacts when many others are calling. These and many other skills are very valuable to excelling on the HF bands, and are the key to gaining more enjoyment from operating.

## Aspects of the hobby

There are many different ways that people enjoy the HF bands. One of the most popular is "DXing" which is searching out stations from distant or interesting countries. It is quite possible to make contact with stations in virtually every country in the world and top DXers have country scores in excess of 300. As a further challenge, these countries can be contacted on several bands. People also enjoy making contact with stations on different islands.

DXers soon become highly skilled operators but they also have to ensure their stations are operating at maximum efficiency. This means that they spend a lot of time improving their equipment, both inside and outside the shack.

At certain times of the year, the amateur radio bands spring to life with an enormous amount of activity as a result of a contest. Amateur radio contests take a variety of forms but usually involve participants making as many contacts as possible, often with stations in a particular continent or country. They are great fun to join, and they also give amateurs the opportunity to contact stations from rare countries.

# HF AMATEUR RADIO

Photo 1.1: Station of a top UK DXer

Some countries have little or no amateur radio activity and expeditions are often launched to activate them. Not only is this great fun for those operating the station, but it also gives people the chance to visit some far-away and interesting places. In many cases, these DXpeditions are organised to coincide with one of the major contests to ensure they make the maximum number of contacts.

Once a contact have been made, it is common for radio amateurs to exchange QSL cards to confirm the contact. Collecting these from stations around the world can be a fascinating pastime of its own. In addition, some people like focusing on gaining operating awards. Achieving one of these awards can give an amateur a great sense of achievement.

Many operators enjoy talking to old friends and the HF bands can provide the medium for keeping up with them, even if they live many hundreds or even thousands of miles away. It is not unusual for radio amateurs to chat with overseas friends once a week or more using the HF bands.

There is plenty of scope for experimentation on the HF bands. Antennas are a particularly interesting area because even small improvements to an antenna can greatly enhance the performance of a station. Time and energy spent on improving an antenna is always worthwhile and can help the operator contact an elusive or rare station. However, there are lots of other opportunities for experiments. Many people like to build their own equipment. Although most could not hope to produce a commercial standard receiver or transceiver, there is still plenty of scope for the home constructor. There are lots of useful ancillary items that can be built relatively easily. Also there is a growing number of QRP (low-power) operators who use simple transmitters and receivers that they built themselves. Not only is building your own equipment a fun challenge, but it also develops operating skills because of the low powers that are often used.

There is a wide range of transmission modes that can be used. Single sideband is almost exclusively employed for speech transmissions. However,

# CHAPTER 1: AN INTRODUCTION TO THE HF BANDS

**Photo 1.2: The D68C expedition to the Comoros Islands achieved 160,000 contacts. Many of these contacts were made with people who had average or modest stations**

Morse (CW) is very popular. It might come as a surprise but Morse still has technical advantages over other transmission modes and can succeed in making contacts where other modes would fail. Not only this, many people really enjoy using it too. There are also data modes like radio teletype and the more

# HF AMATEUR RADIO

Photo 1.3: A selection of QSL cards and awards

up-to-date versions like AMTOR and PSK31. Slow-scan television is also used and provides a considerable amount of interest for some.

HF operators can also help advance communication technology. In the early days of amateur radio, it was radio amateurs who discovered the value of the short-wave bands, and even today much work is being undertaken to improve understanding of how signals propagate. Meanwhile, radio amateurs are developing new forms of data transmission to enable data to be transmitted more accurately through interference or to give more facilities.

Amateur radio can be used to help the community, especially after disasters. Often amateur radio provides the only working communications link for small islands that have been hit by storms. Similarly amateur radio has provided vital emergency communications after earthquakes and floods. The resilience and ingenuity of amateur operators under these circumstances has saved countless lives. Although the UK is rarely hit by hurricanes, amateur radio has nevertheless been used on many occasions to help those in distress.

The radio frequency spectrum extends over a vast range. It is split up into

| Table 1.1: UK HF Amateur Radio Bands | |
|---|---|
| Frequencies (MHz) | Band name |
| {0.1357 – 0.1378137 kHz} | not strictly an HF band |
| 1.810 - 2.000 | 160 Metres (Top Band) |
| 3.500 - 3.800 | 80 Metres |
| [3.500 - 4.000 | 75 Metres - USA] |
| 7.000 - 7.200 | 40 Metres |
| [7.000 - 7.300 | 40 Metres - USA] |
| 10.100 - 10.150 | 30 Metres |
| 14.000 - 14.350 | 20 Metres |
| 18.068 - 18.168 | 17 Metres |
| 21.000 - 21.450 | 15 Metres |
| 24.890 - 24.990 | 12 Metres |
| 28.000 - 29.700 | 10 Metres |

## CHAPTER 1: AN INTRODUCTION TO THE HF BANDS

Photo 1.4: Radio amateurs provided communications following the Lockerbie plane crash

various areas to enable sections to be referenced more easily. Strictly speaking, the HF (high-frequency) portion covers 3.0 to 30MHz and includes those frequencies that are referred to as the short-wave bands. However, the short-wave bands may be considered as any bands having a frequency between the medium-wave broadcast band and 30MHz.

Within these limits, there are a number of bands allocated to amateur radio. 'Top Band', covering 1.81 to 2.0MHz in the UK, is is treated as a short-wave band, although not strictly part of the HF section of the spectrum.

Photo 1.5: Amateur radio proved to be the only form of communication available when a hurricane hit Samoa

5

HF AMATEUR RADIO

CHAPTER 2

# Radio Wave Propagation

*In this chapter:*
- Ground waves and sky waves
- The atmosphere
- The ionosphere
- Regions in the ionosphere
- Variations in the ionosphere
- Sunspots and sunspot cycle
- Definitions and terms
- Path losses and fading
- Multiple hops
- Ionospheric storms
- Grey-line propagation
- Predicting conditions

ONE of the key skills available to any HF DXer is a knowledge of radio signal propagation. Knowing when to listen, which frequencies to use and where the signals may be coming from give the experienced DXer a vital edge. A good knowledge of propagation is an essential weapon in any DXer's armoury. Propagation is also a fascinating topic, and many people like to study it. Every operator should have at least a basic grasp of propagation, in order to make contacts with the stations and countries they want.

There are many ways in which signals propagate. The most obvious is when they they travel in free space. In this case, the signal spreads out in all directions like the ripples on a pond after a stone is dropped into the water. However, when signals are transmitted from a station on Earth, they are affected by the close proximity of the ground as well as other elements, including the ionosphere. In fact, most signals on the MF and HF bands are heard via what is termed the ground wave, or after they have been refracted back to ground from the ionosphere via what is termed the sky wave. As a result, signals are able to travel further than the line-of-sight distance, and often as far as the other side of the globe.

## Ground waves and sky waves

Ground waves occur as the signal spreads out from the transmitter. Instead of traveling in a straight line and not being heard beyond the horizon, the signal tends to follow the curvature of the Earth. The reason for this is that currents are induced in the surface of the Earth, slowing the wavefront close to it. This

# HF AMATEUR RADIO

has the effect of tilting the wavefront downwards so that it follows the Earth's curvature and can be heard beyond the horizon.

Ground-wave signals become attenuated more at higher frequencies and coverage is reduced. As a result, it is not generally used for signals much above 2 or 3MHz.

A coverage area extending to 150km and more may be expected for a high-power broadcast station operating on the medium waveband, whereas a short-wave broadcast station will have a very small coverage area. Short-wave stations (including radio amateurs) rely on the sky-wave signals that travel away from the Earth and towards the ionosphere to give ranges of many thousands of miles.

## The atmosphere

Before taking a look at how signals are reflected by the ionosphere, it is useful to find out about the areas where reflections take place and how these areas are formed. The atmosphere can be split up into a series of different layers according to their properties.

The area closest to the Earth is known as the troposphere. This extends to an altitude of about 10km and it has an effect on propagation mainly in the VHF and UHF bands. It does not noticeably affect the short-wave bands. Above the troposphere is an area known as the stratosphere. This is found at altitudes between about 10 and 50km and contains the famous ozone layer at an altitude of about 20km. Next is the mesosphere, extending to an altitude of about 80km, and then on top of this is the thermosphere where temperatures can soar to 1200°C.

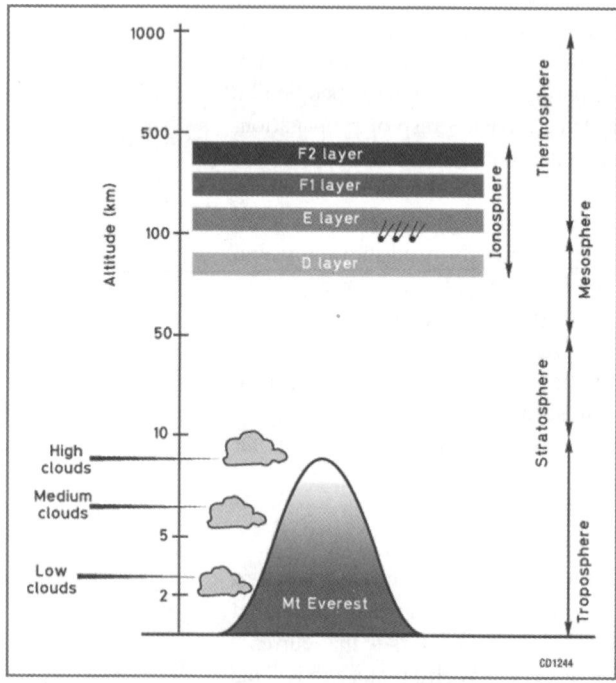

Fig 2.1: The composition of the atmosphere

For short-wave communications, an area known as the ionosphere is all-important. This crosses several of the meteorological boundaries and extends from an altitude of about 50km right up to about 600km.

## The ionosphere

The ionosphere is named so because it is where chemical ions exist. In most areas of the atmosphere, gas molecules are in a combined state and are electrically neutral. However, in the ionosphere, gas molecules become ionised, forming a positive ion (a molecule that has lost an electron) and a free electron – it is actually these free electrons and not the positive ions that affect radio waves.

Ionisation occurs as a result of intense solar radiation at these altitudes splitting the gas molecules. Of all the forms of radiation from the Sun, it is mainly the ultraviolet light that causes this effect. Ionisation can first be detected at an altitude of around 30km, but the electron density is not high enough to affect radio signals until an altitude of around 50km is reached.

The ionosphere is often thought of as a number of distinct layers. The different layers can be visualised as peaks in the level of ionisation. There are three such peaks: the lowest is known as the D layer, above this is the E layer and still higher is the F layer. (Strictly speaking, there is also a C layer but the degree of ionisation there is so low that it has no effect on radio waves.)

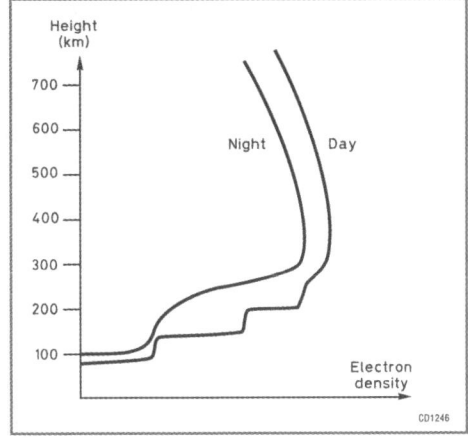

Fig 2.2: Typical electron distribution in the ionosphere

## The D layer

The D layer is found at altitudes between 50 and 80km. It is only present during the day when the Sun's radiation is present. The reason for this is that free electrons and positive ions recombine to form neutral molecules. At this altitude, the air density is relatively high and recombination occurs quite quickly – as a result, the Sun's radiation is needed to retain the level of ionisation. When the radiation is removed, the level of ionisation quickly falls and the layer effectively disappears at night.

Fig 2.3: Variations in the ionosphere over the period of a day

The D layer acts as an attenuator to radio signals, and this is particularly noticeable at low frequencies. This can be seen by the fact that signals in the medium-wave broadcast band are prevented from reaching the higher layers during the day and they are not heard beyond the range of the ground wave. The level of attenuation varies according to an inverse square law. In other words, if the frequency doubles, the level of attenuation falls by a factor of four.

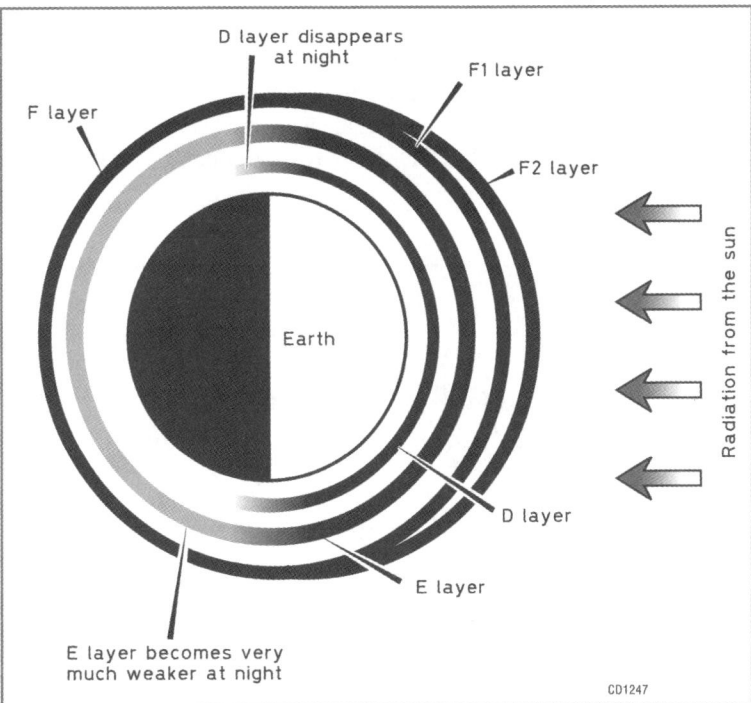

The signals are attenuated because the electrons in the layer are vibrating in synchronism with the frequency of the signal. The air

9

density is still relatively high at this altitude and the vibrating electrons collide with other air molecules. A small amount of energy is lost at each collision and the signal is reduced.

The actual level of attenuation is related to the number of collisions that take place. This is obviously reliant on the level of ionisation but it also depends on the frequency of the radio signal. As the frequency increases, so the wavelength of the vibrations decreases and the number of collisions falls.

As a result, low-frequency signals are attenuated more than higher-frequency ones, although it should be remembered that high-frequency signals still suffer some attenuation.

## The E layer

The E layer appears above the D layer and can be found at altitudes between about 100 and 125km. Here, too, the ions combine relatively quickly and after dark the level of ionisation falls quite rapidly, although a small amount of residual ionisation remains at night.

When signals enter the E layer, they cause the electrons to vibrate. However, the air density is much less than at the altitude of the D layer and there are far fewer collisions. As a result, much less energy is lost and the layer affects radio waves in a different way. Unlike the D layer, where the radio signal causes the electrons to vibrate and collide with other gas molecules, the lower gas density means that when the electrons vibrate, the signal is actually re-radiated. Because the signal is traveling into an area where the level of ionisation (and hence the number of free electrons) is increasing, it is bent or refracted away from the area of higher electron density. This refraction is often sufficient at frequencies in the HF bands to bend the signal back to Earth, as if it had been reflected.

These 'reflections' are affected by the frequency of the signal and the angle at which the signal enters the layer. As the frequency increases, so the amount of refraction decreases and a point is reached where the signals pass straight through.

Fig 2.4: Ionospheric signal propagation at different frequencies

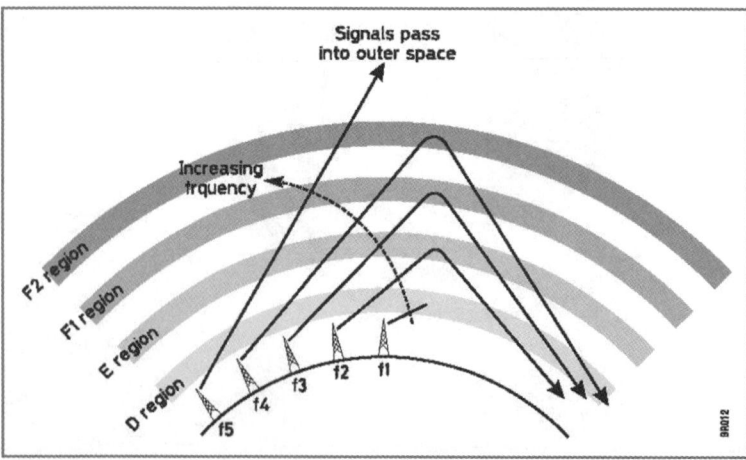

## The F layer

The highest and most important of the ionospheric layers for long-distance communication is the F layer. During the day, it often splits into two sub-layers that are called the F1 and F2 layers. Then at night they merge back into a single layer.

The height of the layers varies considerably according to the time of day, the season and the state of the Sun. However, as a rough guide, in summer the F1 layer may be at an altitude centred around 300km and the F2 layer centred around 400km. In winter these figures may fall to around 200 and 300km. At night, the combined F layer is generally between about 250 and 300km.

The level of ionisation of the F layer falls just like the other layers but, because the level of ionisation is much greater, there is still sufficient ionisation present to affect signals at night. Like the E layer, the F layer acts as a reflector of signals rather than an attenuator. Also, it is the highest of the layers and so the distance that can be achieved using a single reflection from it is the greatest.

## Varying the frequency

During the day, signals on the medium-wave band are only able to use ground-wave propagation because the D layer absorbs any sky-wave signals. At night, when this layer disappears, signals can be reflected back to Earth.

If the frequency of the signal is increased, the attenuation introduced by the D layer reduces and a point is reached when the signal can pass though it and on to the E layer above. Here the signal will be reflected and pass back to Earth. However, as the frequency increases, the degree of refraction becomes less, and the signal will penetrate further into the layer until a point is reached where the signal passes through the E layer and onto the F layer. As the frequency increases still further, the process is repeated for either the F layer or the F1 and F2 layers, and eventually a point is reached where the signal passes through all the layers and travels on into outer space.

## Variations in the ionosphere

The ionosphere is constantly changing and this results in changing conditions on the HF bands. one factor that affects the ionosphere is the amount of radiation received from the Sun, and hence the time of day is a major factor. Similarly, the season has a major effect – in the same way that more heat is received from the Sun in summer, more radiation hits the ionosphere. Furthermore, the actual state of the Sun is very important, in particular the number of sun-spots on its surface. Sunspots affect the ionosphere because the areas around the spots emit greater amounts of ultraviolet light - one of the main causes of ionisation.

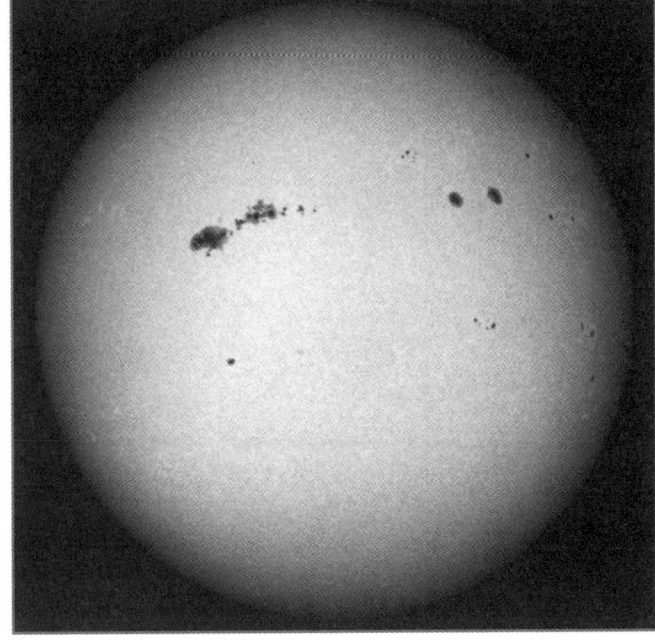

**Photo 2.1: The Sun, showing several sunspots**

11

The prevalence of sunspots varies cyclically with a period of around 11 years. As a result, ionospheric conditions and hence radio propagation also vary in line with this cycle. Broadly speaking, at the low point of the cycle, the HF bands above about 20MHz may not support ionospheric propagation, while at or near the peak of the cycle, frequencies of 50MHz and higher may be affected.

## Sunspots and sunspot cycle

Even though sunspots have a major effect of ionospheric radio propagation conditions and a number of other geophysical phenomena, the variation in their numbers was not understood for many years. Their presence had been observed on the surface of the Sun for many centuries, but no pattern had been seen. The reason for this was that the daily numbers varied very widely.

In order to view the trends, the data needs to be averaged or smoothed over a longer period. A two stage process is usually used. First, the daily numbers are averaged over the period of a month and then the monthly figures are smoothed over a 12 month period. In order to ensure that the mean falls right in the middle of the month in question rather than between months, the period of smoothing is run over 13 months, but taking half the value for the months at either end

Both the monthly average sunspot numbers and the smoothed values are available for use in propagation predictions, although the smoothed figures are much in arrears. The numbers are now prepared by the Sunspot Index Data Centre in Brussels from information supplied by a number of observatories. They appear in DX propagation information available from a wide variety of sources including the RSGB. The 12-month smoothed sunspot number correlates quite closely with the prevailing HF radio propagation conditions.

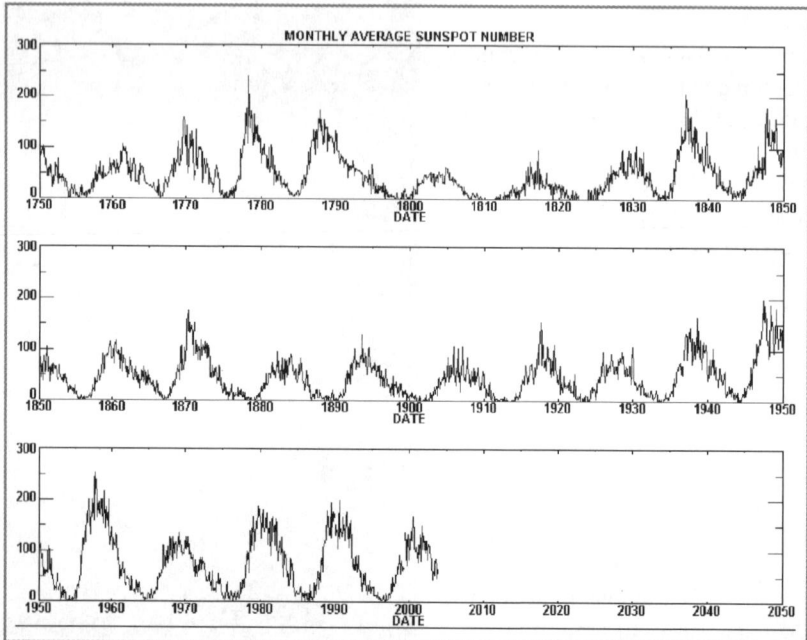

Fig 2.5: The sunspot cycle since records began (Image courtesy of NASA)

Once the numbers had been smoothed, a trend was detected and it could be seen that there was cyclical variation with a period of about 11 years. These cycles are named by number, starting with Cycle One which began in 1755. Since then there have been over 20 more, with Cycle 22 ending in the latter part of summer 1996.

From analysis of records, a number of factors have become obvious. The first is that the cycles are by no means very regular. Although the average cycle length (measured as the time between two peaks) is about 10.9 years, the length varies anywhere from just over 7 years to 17 years. The smoothed sunspot numbers also vary widely. The maximum sunspot activity numbers vary from 49 to 200 with an average of just over 100. The minimum number can be anywhere between none and 12.

Usually after the sunspot minimum, sunspot activity rises sharply, reaching a peak in around four years, and after this it falls away more slowly, taking around seven years to decay. Naturally, this figure too varies very widely and it can only be taken as a very rough guide.

The sunspot cycle is of great interest to anyone using the HF portion of the radio spectrum. Propagation conditions are greatly influenced by sunspot activity, and accordingly they vary in line with the sunspot cycle. At the low point of the cycle, the high frequency bands above 20MHz or so may not support ionospheric reflections, whereas at the peak of the cycle frequencies at 50MHz and higher may be reflected.

## Ionospheric propagation definitions and terms

When talking about the ionosphere there are a number of terms that are commonly used. The first is a signal's angle of radiation. This is effectively the angle between the main beam of the signal and the ground. Very low angles of radiation travel almost parallel to the Earth at first. High angles of radiation travel upwards towards the ionosphere as shown in Fig 2.6 and hit it with a high angle of incidence. They will need to receive a higher degree of refraction to be returned to the Earth, and are therefore more likely to pass through

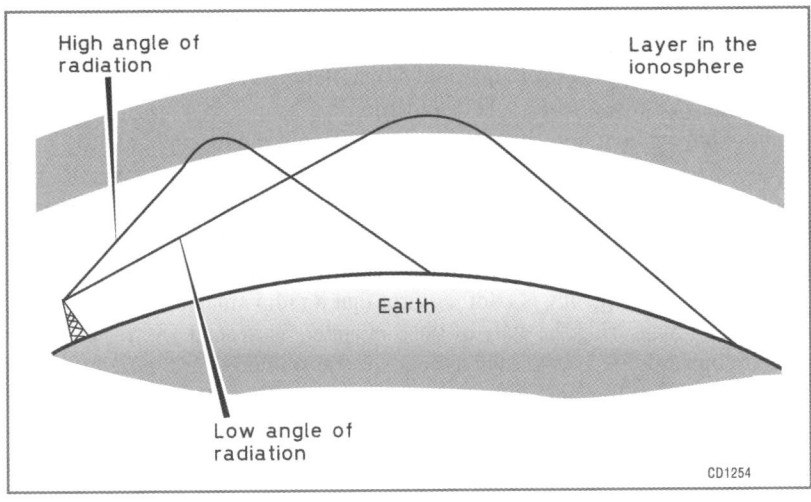

Fig 2.6: The angle of radiation, showing how low-angle signals travel further

# HF AMATEUR RADIO

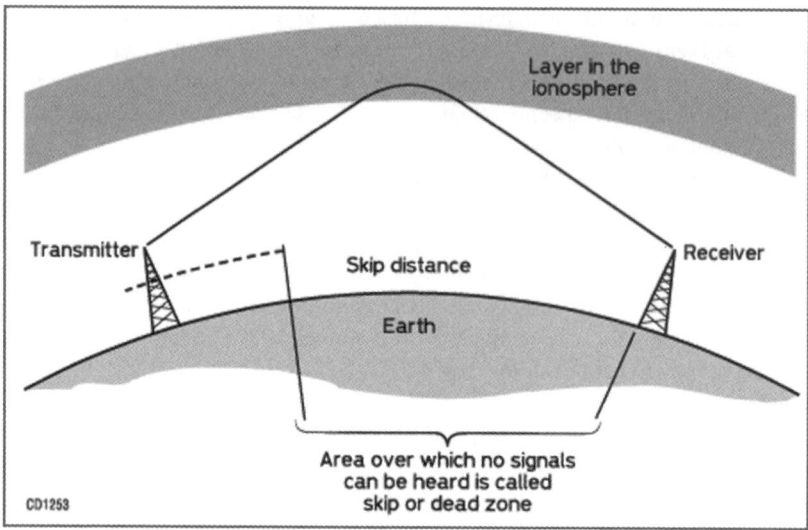

Fig 2.7: Skip distance and dead zone

a layer than signals with a lower angle of incidence. Additionally, signals with a low angle of radiation are able to travel further as a result of the geometry.

Another term is skip distance. This is the distance a signal travels along the surface of the Earth as shown in Fig 2.7. The maximum distance that can be achieved using the E layer is generally considered to be 2000km (1,250 miles), but this is reduced to just 400 km (250 miles) if the angle is 20 degrees. Similarly, the maximum distance achievable using the F2 layer reduces from around 4000km (2,500 miles) to just under 1000km (600 miles).

It is worth noting at this point that to be able to place the maximum amount of energy where it is required, it is necessary to have a directional antenna. All antennas radiate more energy in some directions than others. Their radiation pattern can be plotted out in what is termed a polar diagram. To ensure the optimum performance, the antenna should be orientated in the correct direction, and it should have the correct angle of radiation. In many applications where the maximum distance is required, an antenna with a low angle of radiation is needed. Having said that, particularly for low frequencies that may be affected by the D region attenuation, this means that signals will have to travel through the D region for longer and will suffer greater levels of loss. Not all applications require a low angle of radiation.

Looking again at how a signal is reflected from the ionosphere in Fig 2.7, it can be seen that there is a region in which no signal may be heard. This occurs in the region after the ground wave has been attenuated to the extent it cannot be heard, and before the first sky-wave signals are returned to Earth. This is known as the skip zone or dead zone.

The critical frequency is another term that a radio amateur will encounter. One way of measuring the state of the ionosphere is to send a series of pulses directly upwards. This is known as ionospheric sounding. If the frequency of the transmitted pulses are gradually increased, it is found that at first the pulses are returned to the Earth. As the frequency is increased, the signal penetrates further into the layer and eventually a point comes where it passes through it. The point at which the signal just passes through a layer and on to the next is called the critical frequency for that layer. Also, by measuring the time taken for a pulse to be returned it is possible to measure the effective height of a given layer.

Another term, the lowest usable frequency (LUF), is very important. As the frequency of a transmission is reduced, the losses increase, and a point is reached where the signal cannot be copied. The LUF is defined as the frequency where the signal equals the minimum strength for satisfactory reception.

There is also the maximum usable frequency (MUF), which is more relevant to the other end of the spectrum. As the frequency of a signal is increased, the signal penetrates further into the layers and eventually passes right through. A point is reached where communications just start to fail. This is the maximum usable frequency. Generally, the MUF is between three and five times the critical frequency.

## Path losses

As a signal travels over any path, its strength is reduced for several reasons. The major one that has already been mentioned is the loss arising from the D region and in some instances from the lower regions of the E region.

Loss also arises as the signal spreads out as it travels, and the area covered by the wave fronts on the signal increase in area and accordingly the signal decreases in intensity.

It might be expected that the distance travelled by a signal is the great circle distance between the transmitting and receiving stations. However, this is not exactly the case because the signal does not follow the curvature of the Earth, but travels up to the ionosphere and then returns downwards again. As a result, the distance travelled is larger than the great circle distance. Over paths where the angle of radiation is very low, the difference may be acceptably small, but over other paths where the angle of radiation from the antenna is much higher, there can be a significant difference. Accordingly, higher levels of loss than expected might be incurred.

Further losses occur when the signal's polarisation is changed by the ionosphere. Even though signals that enter the ionosphere from terrestrial antennas are normally linearly polarised, the action of the ionosphere with the Earth's magnetic field causes them to be elliptically polarised. Normally, this loss is grouped together with a variety of other small losses. The extent of this loss varies according to a number of factors including the geomagnetic latitude, the season, time of day and the length of the signal path. Typically, this loss is around 9 dB.

## Fading

Fading and signal variations are major consequences of ionospheric propagation. The signal variation may be fairly shallow, with the signal changing in level between 10 and 20 dB, or it may cause the signal to completely disappear. There are a number of reasons for this but they all result from the ever changing state of the ionosphere.

# HF AMATEUR RADIO

One of the major causes of fading is multi-path interference. Even with directive antennas, the signal will illuminate a wide area of the ionosphere. As it is very irregular, the signal will reach the receiving station via a number of paths, each with different path lengths. The overall received signal is the summation of them all. As the ionosphere changes, the signals will fall in and out of phase with one another, resulting in a considerable variation in the strength.

This may also be noticed on MF signals at night. Normally signals are audible via the ground-wave during the day, but at night the sky-wave may also be audible. As the state of the ionosphere changes, so will the path length of the reflected signal, and accordingly the phase will vary. This will give rise to fading of the overall signal as the ground-wave and sky-wave signals interfere.

Irregularities in the ionosphere may also cause the path lengths of closely spaced frequencies to be different. For signals such as amplitude modulation (AM) and single sideband (SSB), this may result in a reduction in some frequencies in the audio range, while others are intensified. When this "selective fading" occurs, the result can be serious distortion of an AM signal if the carrier suffers the selective fading and a reduction in level. Since SSB signals do not depend on the transmission of a carrier, this mode is less affected by this form of fading. Additional synchronous detection of an AM signal also significantly improves signal quality under these conditions.

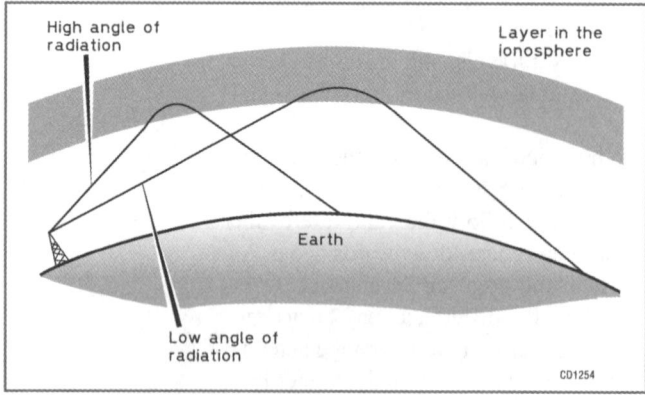

Fig 2.8: A signal is received via several paths and as changes occur in the ionosphere signal lengths vary, causing signals to fall in and out of phase with one another. As a result, the overall signal strength varies with time.

The significant variation in free electron density in the ionosphere gives rise to other forms of fading. The ability of a region of the ionosphere to reflect a signal at a given frequency may change. When operating near the MUF, the signal may fade as the signal starts to pass through the region.

There may also be regions or clouds of high electron density in the D region. Accordingly, as the clouds move, signals, especially those lower in frequency, will fade as the cloud passes through the signal path. Fading of this type normally takes place over a period of half an hour to an hour and may reduce the signal by a figure between 5 to 10dB.

A further type of fading occurs when the ionosphere causes changes in the polarisation of the signal. As the polarisation of the incoming signal changes, so will the signal strength of the signal picked up by the antennas as it falls more closely in line with the antenna polarisation and then away from it.

## Multiple reflections

The maximum distance that can be achieved when using a reflection from the F2 region is around 4000km and for the other reflective regions in the ionosphere it is somewhat less  By simply using a reflection from the ionosphere, it is not possible to explain how signals travel from the other side of the globe using ionospheric propagation. The reason this is possible is that signals undergo several reflections. It is found that after the signals return to Earth from the ionosphere they are reflected by the Earth's surface and returned back up to the ionosphere. Here they are reflected again by the ionosphere, being returned to the Earth a second time about twice the distance away from the transmitter that a single reflection would give.

Unfortunately the signal undergoes additional attenuation in this process. Each reflection by the Earth introduces some losses and therefore the signals are attenuated each time they are reflected. The surface of the Earth at the point of reflection has a major effect on the level of these losses. Sea water is a very good reflector, but dry desert is very poor. This means that signals that are reflected in the Atlantic are likely to be stronger than those reflected by a desert region.

Apart from the reflection at the Earth's surface, the signal suffers losses in the ionosphere as well. Every time the signal passes through the D region, there is additional attenuation. This can be very important because the signal has to pass through the D layer twice each time it is reflected by one of the higher regions and with more than one hop, the signal passes through the D region several times. As already mentioned, the attenuation reduces with frequency. Apart from the fact that high frequency paths are more likely to use the F2 layer and have less reflections, the high frequency path will also suffer less loss from the D layer. This means that a signal on 28MHz, for example, will be stronger than one on 14 MHz.

It should also be remembered that the path length for a multiple reflection signal will be greater than the great circle distance around the globe, especially if high angles of radiation are used. This in itself will add to the signal loss because the loss is proportional to the path length.

## Chordal hop

At some times, a tilt in the ionospheric regions and in particular the F2 region may occur. When this happens, the signal may not be reflected back to Earth. Instead it is reflected to another part of the ionosphere, before being reflected back to earth. As this form of propagation does not involve a reflection from the surface of the earth, the losses are much lower and accordingly signal strengths provided are higher. In some research, it has been proposed that this form of propagation could account for round-the-world echoes.

The tilt or distortion in the ionosphere required to produce this form of propagation is known as Chordal Hop. It occurs near sunrise and sunset and across the equator. The propagation using the equatorial anomaly generally

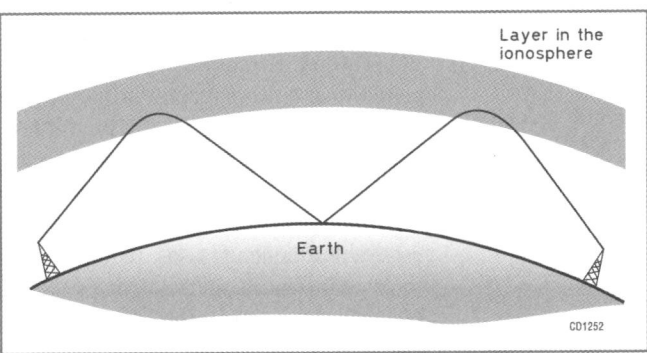

**Fig 2.9: Multiple reflections from the ionosphere**

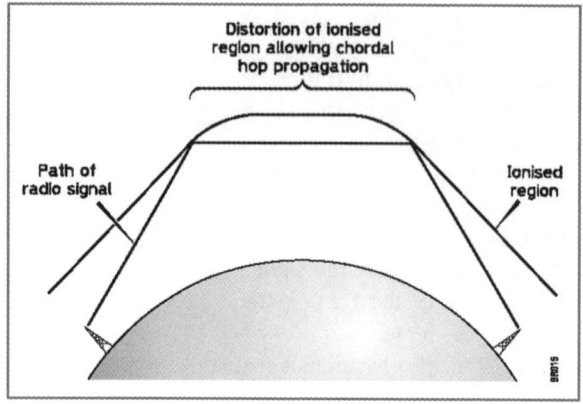

Fig 2.10: The north-south chordal signal path across the magnetic equator

occurs in a north-south (or south-north direction). It is found that the F2 region is higher across the equator and this means that either side of this it tilts, enabling the signals to be reflected above the earth's surface.

## Ionospheric disturbances and storms

On occasions, normal propagation on the HF bands can be disrupted. Sometimes this may be for a few hours, but at other times it may last for a few days. There are a number of reasons for this, but one of the main causes is a solar disturbance. There are a number of types of disturbance that can occur on the Sun, some of which are only starting to be understood. When a disturbance occurs, huge amounts of material may erupt from under the surface of the Sun and be thrown out into space. Some of the particles emitted travel at a colossal velocity – about a tenth the speed of light, while others are significantly slower, traveling at only about 1000km per second. Along with the eruption, vast amounts of radiation are also emitted. When seen using the right equipment, the flare looks like an enormous flame leaving the surface of the Sun. These events can last for about an hour, after which the sun's surface settles back to its former state.

Disturbances have a number of effects on radio communications. One is seen when the radiation arrives. The level of ionisation in the D layer rises sharply, often absorbing signals throughout the HF spectrum, although this only occurs on the sunlight side of the Earth. This effect is known as a sudden ionospheric disturbance (SID) and may last for a few hours. Depending on its severity, it may only affect the lower frequencies.

It takes a few hours for the high-energy particles to arrive. When they do, they are deflected by the Earth's magnetic field so that they move towards the poles where they cause a very large increase in the level of D-layer absorption. This can last for up to three or four days, and during this time such polar cap absorption can prevent HF communications across the poles.

Between 20 and 40 hours later, the lower-energy particles arrive. These cause a change in the Earth's magnetic activity and this is recorded by changes in the A and K indices which act as an indicator of the stability of HF band conditions. A geomagnetic storm may give rise to an ionospheric storm. If this occurs, the chemistry of the ionosphere is affected, significantly depressing the levels of ionisation and reducing the frequencies that can be reflected by the E and F layers. At the same time, there is an increase in the level of absorption. The total effect is that the maximum usable frequency falls and the lowest usable frequency rises, reducing the band of frequencies that can be used. In some instances, the MUF and LUF may meet and there will be a total radio blackout. In this case, the only stations that can be heard are via ground wave. The effects may last for up to a week for a bad storm.

# CHAPTER 2: RADIO WAVE PROPAGATION

Given that solar disturbances have such a major impact on ionospheric propagation conditions, they are reported in a variety of ways and by a number of different organisations. In general, there are three major adverse effects on the ionosphere: Geomagnetic storms, Solar radiation storms, and Radio blackouts. These correspond to the G, S, and R designations used in the WWV report at 18 minutes past the hour as in table 2.1. These disturbances result from three main elements arising from the disturbances, namely an increase in the level of radiation, very high energy protons and finally an increase in the level of the solar wind.

WWV is a radio frequency standard run by the National Institute of Standards and Technology in the USA. WWV broadcasts time and frequency information 24 hours per day, seven days per week and it is located in Fort Collins, Colorado, about 100km north of Denver. The broadcast information includes time announcements, standard time intervals, standard frequencies, time corrections, a BCD time code, geophysical alerts, marine storm warnings, and global positioning system (GPS) status reports.

The station radiates transmission of 10kW on 5, 10 and 15MHz; and 2.5kW on 2.5 and 20MHz. Each frequency is broadcast from a separate transmitter. Although each frequency carries the same information, multiple frequencies are used because the quality of HF reception depends on many factors such as location, time of year, time of day, the frequency being used, and atmospheric and ionospheric propagation conditions. The variety of frequencies makes it likely that at least one frequency will be usable at all times.

## Grey line propagation

At dawn and dusk, signals are able to travel over greater distances and with much lower losses on some paths and frequencies than might normally be expected. This occurs by what is termed grey line propagation. This happens when signals travel along the grey or twilight zone between night and day. This area where night and day meet is also known as the terminator.

The improved propagation conditions around the grey line occur primarily on the lower frequency parts of HF band. The improvement in propagation is due to a much reduced level of ionisation in the D layer. Even though D layer ionisation is lower, the level of ionisation of the F layer, which gives most of the long distance signal propagation, is still high.

There are two main reasons for the ionisation disparity between the D and F layers. The first is that the level of ionisation in the D region drops very quickly around dusk and after dark because the air density is high and recombination of the free electrons and positive ions occurs comparatively quickly. The second reason is that

Table 2.1: The G, S, and R designators used in the WWV transmissions

| Geomagnetic Storms | Solar Radiation | Storms | Radio Blackouts Descriptor |
|---|---|---|---|
| G1 | S1 | R1 | Minor |
| G2 | S2 | R2 | Moderate |
| G3 | S3 | R3 | Strong |
| G4 | S4 | R4 | Severe |
| G5 | S5 | R5 | Extreme |

the F layer is much higher in altitude, and as the Sun sets it remains illuminated by the Sun's radiation for longer than the D region, which is lower down. Similarly, in the morning when the sun is rising, the F region receives radiation from the Sun before the D region and its ionisation level starts to rise before that of the D region. As the level of the D region ionisation is low, this means that attenuation of lower frequency signals is very much less than in the day. This also occurs at a time when the F region ionisation is still very high, and good reflections are still achievable. Accordingly, this results in much lower overall path losses around the grey line than are normally seen.

It should be remembered however that the radio terminator does not exactly follow the day time/night time terminator on the Earth's surface. The ionised regions are well above the Earth's surface and are accordingly illuminated for longer, although against this the Sun is low in the sky and the level of ionisation is low. Furthermore, there is a finite time required for the level of ionisation to rise and decay. As there are many variables associated with the "radio signal propagation" terminator, the ordinary terminator should only be taken as a rough guide for radio signal propagation conditions.

Frequencies that are affected by this form of propagation are generally limited to around 10MHz. This means that the bands primarily affected are 80 (75), 30 and 30m. Frequencies higher in frequency are attenuated to only a minor degree by the D region and therefore there is little or no enhancement.

It is, however, still possible for higher frequency signals to be affected by a grey line type enhancement. This occurs when a propagation path is opening in one area and closing in another giving a short window during which the path is open on a particular frequency or band of frequencies.

Looking at the MUFs over the course of the day can be a useful way of understanding this phenomenon. The level of ionisation in the F layer falls after dusk, and rises at dawn. This results in the MUF falling after dark. Accordingly, stations experiencing dawn find that the MUF rises and those experiencing dusk find it that it falls. For frequencies that are above the night time MUF, and for stations where one is experiencing dusk and the other dawn, there is only a limited time where the path will remain open. This results in a similar effect to the lower frequency grey-line enhancement.

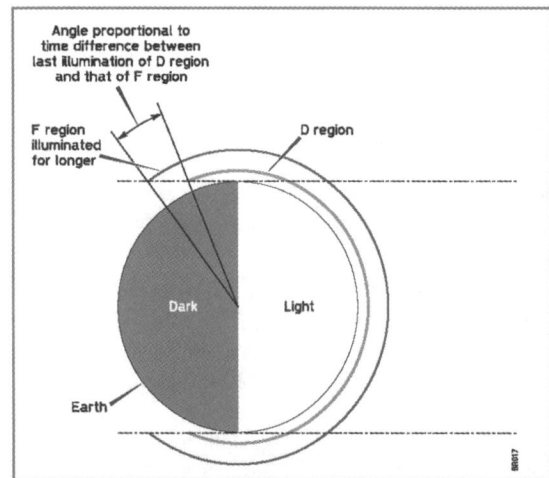

Fig 2.11: The F region remains illuminated for longer than the D region

The path of the grey line changes during the course of the year. As the angle subtended by the Sun's rays changes with the seasons, so the line taken by the terminator changes. This is because during the winter months, the Northern Hemisphere of the earth is titled away from the Sun, while it is tilted towards it during the summer months. The converse is obviously true for the southern hemisphere. In addition, the width of the grey line also changes. It is much wider towards the poles because the line between dark and light is less well defined because the Sun never rises high in the sky at the poles. It is also narrower at the equator. This results in grey line propagation being active for longer at the poles than at the equator.

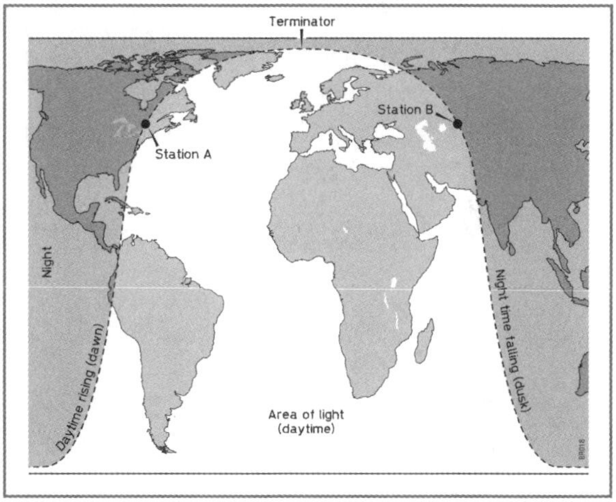

## Other Dusk and dawn enhancements

A number of other enhancements have been noted at dusk and dawn, although they are less well documented and investigated. At times, an enhancement of around 10dB has been noted at paths that cross the terminator line at 90 degrees. Additionally, just as a band is closing, or opening, far fewer stations are likely to be using it, and this fact can be used to advantage in many instances.

Fig 2.12: Grey line like propagation conditions can be seen at higher frequencies as a propagation window opens between Station A and Station B

## Predicting conditions

Knowing how propagation varies on the different bands is one of the key skills for anyone interested in the HF bands. In many respects, predicting propagation is akin to weather forecasting.

As already mentioned, propagation is affected by a number of factors – frequency, time of day, the season, position in the sunspot cycle, and general state of the Sun all affect the way in which the signals can propagate.

The time of day has a significant effect. At night, when no radiation is received from the Sun, the level of ionisation reduces. On low frequencies this means that signals can reach the reflecting layers and they can be heard over much greater distances. However, at higher frequencies the reduction in the level of ionisation means that signals at frequencies that would have been reflected may pass right through the ionosphere. In other words the maximum usable frequency is reduced.

Similar effects occur at the different seasons. In winter the level of radiation received in a particular hemisphere is reduced in the same way that the Sun's rays do not provide as much warmth. This means that the levels of ionisation are reduced. The low-frequency bands open earlier and perform better, the maximum usable frequencies are less, and the high-frequency bands close earlier in the evening.

The sunspot cycle also has an effect. At the low point of the cycle, frequencies below 20 MHz may not support propagation via the ionosphere whereas at the peak of the cycle frequencies in excess of 50MHz may be reflected.

To be able to estimate what the conditions may be like, figures are available that indicate the effects of the sun. One such figure is the solar flux. This is an indication of the amount of radiation being received from the sun. The sun not only emits vast quantities of heat and ultraviolet radiation but also radio energy. The solar flux is measured at a frequency of 2800MHz (10.8cm) and is given in Solar Flux Units (SFU). The actual measurement is made at the

# HF AMATEUR RADIO

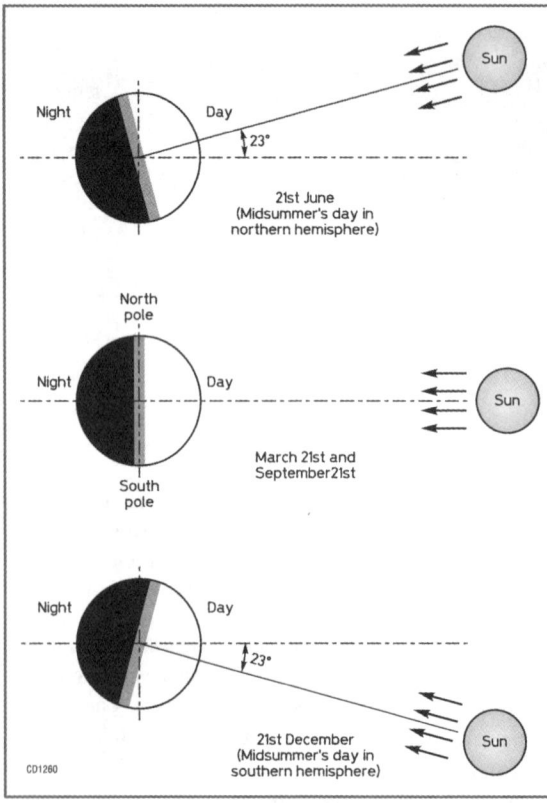

Fig 2.13: The grey line changes with the season

Penticton Radio Observatory in British Columbia, Canada at 1700UTC each day. The value can vary from around 60 up to 300 and beyond. The higher the value, the better the likelihood of good conditions.

Magnetic activity also has an effect on propagation. An index known as the A index reflects the severity of the magnetic flux occurring at local magnetic monitoring points around the globe. During magnetic storms, the index may reach levels of 100 and during severe storms the values may rise as high as 200 and more. As the A index varies from one point on the globe to another an index known as the Ap index is usually quoted - this is the "planetary" index.

The K index is more often used and ranges from 0 to 9. The A index has a 24 hour format whilst the K index is updated every three hours. It is related to the A index as shown in Table 2.1 which also shows the levels of the associated storms.

As a rough guide, the solar flux should be at around 150 to produce good conditions on the HF bands. It should be noted however that it can take several days of high solar flux levels to improve the conditions. Good HF conditions also require the geomagnetic activity to be low with K index levels of between 0 and 2.

As well as using these indexes, HF operators can also use propagation software. Some of these programs are very sophisticated, having been originally developed by broadcast or military organisations. Programs include ASAPS, VOACAP, and many others. Many of them are now available cheaply and can be used on a PC in the shack. It is not intended to review the programs here as they are constantly changing and being updated. Many may also come as "free" software given away with radio magazines, giving HF operators an ideal chance to evaluate the most convenient one for a given situation.

## Further reading:

*Radio Propagation Principles and Practice*, Ian Poole, RSGB, 2004.

Table 2.3: Geomagnetic activity indices

| Conditions | A Index | Potential of a storm |
|---|---|---|
| Quiet | 0-7 | Low |
| Unsettled | 8-15 | Low |
| Active | 16-29 | Moderate |
| Minor storm | 30-49 | Moderate |
| Major storm | 50-99 | High |
| Severe storm | >100 | Very high |

Table 2.2: Conversion between A and K Indices

| K Index | A Index |
|---|---|
| 0 | 0 |
| 1 | 3 |
| 2 | 7 |
| 3 | 15 |
| 4 | 27 |
| 5 | 48 |
| 6 | 80 |
| 7 | 140 |
| 8 | 240 |
| 9 | 400 |

CHAPTER 3

# Types of Transmission

*In this chapter:*

- Morse
- Amplitude modulation
- Single sideband
- Frequency modulation
- Radio teletype
- AMTOR
- PSK31
- Slow-scan television

DIFFERENT types of transmission can be heard on the HF bands, both inside and outside the amateur bands. A few of these transmissions may be intelligible, carrying speech or music, whereas others may be Morse signals. Some signals may appear to be using another form of keying and others might even sound more like computer data.

The reason for using all these types of transmission is that they enable communication to be made in many different ways. Each type has its own advantages. Within amateur radio, several types of transmission are allowed: Morse, amplitude modulation, radio teletype, data, slow scan television, to name a few. This gives a tremendous amount of variety and flexibility in amateur communications.

A radio signal consists of two main components: the carrier and the modulation. The carrier is a steady-state signal which is then modified (modulated) in the transmitter so that it can carry information. When the resulting signal is received, it has to be demodulated so that the information (modulation) is removed from the carrier. If this information consists of sound waves, it may then be passed through the audio amplifier and on to a loudspeaker or headphones. Alternatively, if it consists of data, the demodulated signal may be passed into a computer for processing.

## Morse

One of the oldest, simplest (but still one of the most effective) forms of modulation is Morse code (CW). It has been used for transmissions since the very earliest days of radio - it used for telegraph communications from the middle of the 19th century onwards. Despite its age, it still has a very important place in today's high-technology radio scene.

# HF AMATEUR RADIO

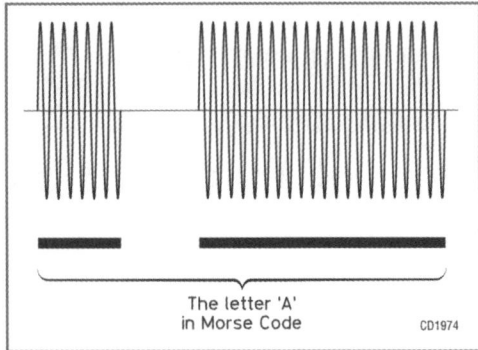

Fig 3.1: A Morse signal as displayed by an oscilloscope

One of Morse's most obvious advantages is that it can be transmitted on very simple equipment. All that is needed is a circuit that generates a radio-frequency signal and a means of turning it on and off. A suitable transmitter can be made from as few as two or three transistors and a handful of other components.

Morse has a number of technical advantages as well. The relatively low modulation rate means that the transmission only occupies a small bandwidth, and as a result the filter bandwidths needed to receive the signal can be made very narrow: 500, 250Hz or less. This is much narrower than the bandwidths of around 2.5kHz required by single sideband speech. Benefits of a narrow bandwidth include less interference from other signals and reduced background noise.

A further advantage is that people can decipher Morse even when it is barely audible because Morse simply consists of a signal that is turned on and off. Morse signals can even be copied when they are below the noise level, whereas a single sideband signal must be above the noise level to be copied. Both these factors give Morse an advantage of more than 10dB (10 times) over other modes. As a result, Morse can be used to make contact when other modes would not succeed, and many stations, especially those with average antennas and transmitter power, use Morse because it enables them to make many more contacts. This is particularly true for those who like to use low power.

Photo 3.1: An old hand key

# CHAPTER 3: TYPES OF TRANSMISSION

However, even those with very good stations use Morse widely to maximise their chances of contacting stations. Morse therefore remains very popular on the amateur bands. You can quickly verify this by listening in the Morse sections of the bands. Indeed, it is often possible to hear Morse stations on bands when no other signals are present.

To resolve a Morse signal, the receiver needs to have a beat frequency oscillator (BFO) as described in Chapter 4. This converts the on-off keying of the carrier into the characteristic note associated with Morse code. All communications receivers and most 'world band' receivers have beat frequency oscillators included, activated from part of the mode switch or from a separate BFO on/off switch. In the absence of a Morse or CW position on a mode switch, the SSB position can be selected since a BFO is also required for resolving this mode.

## Amplitude modulation

Even though Morse possesses many advantages, it cannot convey music or the spoken word. To achieve this, the carrier must be modulated to follow the variations in sound intensity. The simplest method of achieving this is to vary the amplitude of the signal in line with the variations in the audio signal. This form of modulation is known as amplitude modulation (AM) and is widely used for broadcasting on the long-, medium- and short-wave bands. It is also used on the aircraft band above 108MHz but it is rarely used by radio amateurs. The chief advantage of amplitude modulation is its simplicity. Demodulation can be undertaken using a diode and a couple of other components. This means that AM receivers can be made cheaply and easily, as demonstrated by the availability of low cost AM-only radios. However, AM does not use power or spectrum bandwidth very efficiently. Fig 3.2 (a) shows an un-modulated carrier. The instantaneous value of the signal varies (Fig 3.2 (b)) when modulation is applied. The maximum level of modulation that can be achieved is when the envelope falls to zero and rises to twice the steady state value (Fig 3.2 (c)). When this occurs the signal is said to have 100% modulation.

Looking at the frequency spectrum of the signal, it can be seen that if the carrier is

**Photo 3.2: A paddle designed for use with a modern electronic keyer**

25

# HF AMATEUR RADIO

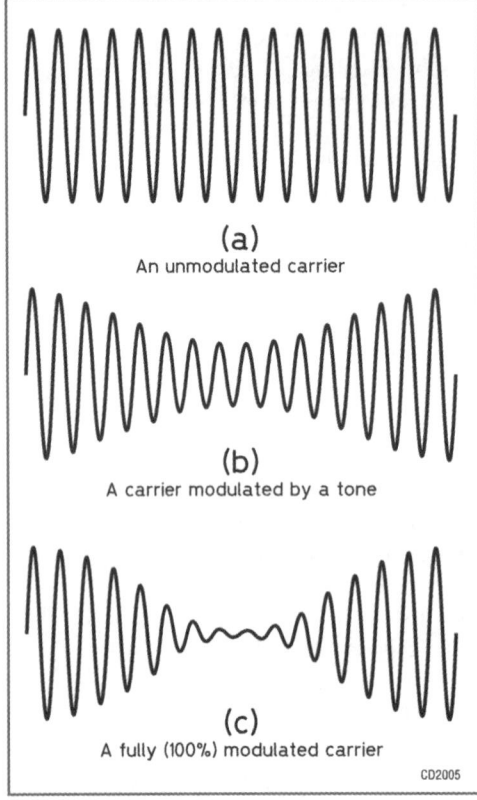

Fig 3.2: Amplitude modulation of a carrier by a sine-wave tone

Fig 3.3: Spectrum of a carrier modulated by a 1kHz tone

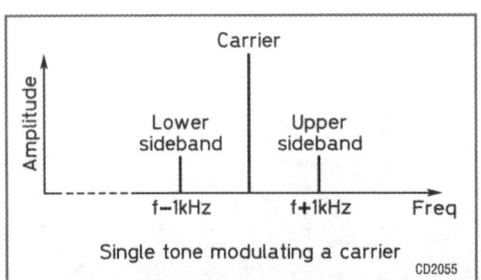

modulated by a single tone, two other signals (sidebands) appear, one on each side of the main carrier (Fig 3.3). A 1kHz tone will produce sidebands spaced 1kHz from the carrier. Under conditions of 100% modulation, the maximum level of the sidebands rises to 50% of that of the main carrier, and this means that the power level of each sideband is only a quarter that of the carrier.

When speech or music is applied instead of a steady tone, sidebands that stretch out either side of the carrier are seen (Fig 3.4). Again the maximum power of these is only a quarter of that of the main carrier. As the actual information for the audio is carried in the sidebands and the carrier only acts as a reference for demodulation, it can be seen that the system is not very efficient. Also, the bandwidth is twice that of the original audio. For these reasons, AM is not used for communications on the HF bands.

## Single sideband

Instead, a type of modulation derived from AM is used by amateurs. Called single sideband (SSB), this modulation sounds garbled when received by an ordinary receiver, but it makes far more efficient use of the available power and can be received at lower signal strengths than AM. In addition, it occupies only half the bandwidth of an AM signal. SSB is used almost exclusively for speech communications on the HF bands, especially by radio amateurs.

The signal is derived from AM by removing the carrier and one of the sidebands (Fig 3.5). This transformation is performed in the transmitter and can be achieved in a number of ways. The most obvious is to use filters but it is also possible to use phasing techniques. Normally, the filter method produces the best results. Here, a fixed-frequency carrier is applied to a double-balanced mixer along with the audio to produce a signal consisting of two sidebands and no carrier. A filter is then used to remove the unwanted sideband. Once this signal is generated, mixers are used to bring the signal to the required frequency.

A BFO and a mixer are required in the receiver for SSB signals to be resolved. Sometimes, the mixer may be called a product detector because the output is the product of the two input voltages. This term was used mainly with valve sets some years ago and is less commonly used these days.

When the carrier is reinserted by the BFO it must be at close to the correct frequency. Any error will manifest itself by raising or lowering the pitch of the reconstituted audio. For amateur communica-

26

tions, a tolerance of around 100Hz can be accommodated, although any offset will give the reconstituted audio a peculiar 'tone'. With most receivers, it is possible to achieve much closer tuning than this and the audio should sound reasonable.

When transmitting or receiving SSB, it is necessary to know which sideband to use: either the upper sideband (USB) or the lower sideband (LSB). The convention for radio amateurs is that the LSB is used on frequencies below 10MHz and the USB above 10MHz.

Many stations use speech processing to improve the quality of SSB transmission. Speech processing can raise the power level of the transmission and make the most effective use of the radiated power by ensuring that only the frequencies that contribute to the intelligibility are transmitted. Full details of these systems are found in Chapter 5 – 'Transmitters'.

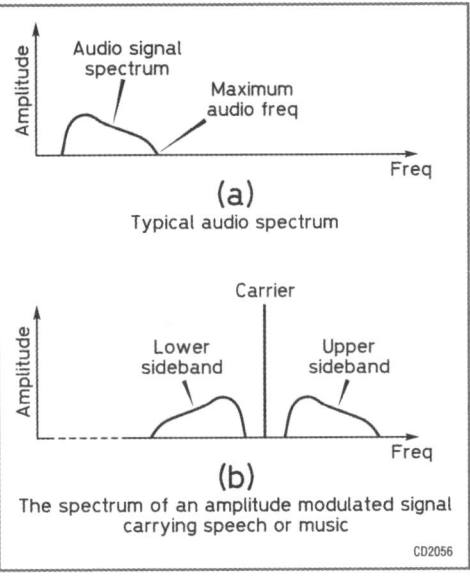

Fig 3.4: Spectrum of a carrier modulated by an audio signal

## Frequency modulation

Although the most obvious way to modulate a signal is to vary its amplitude, there are other methods available to the amateur too. One is to modulate a signal's frequency. The voltage of the modulating signal will determine the exact frequency of the carrier as shown in Fig 3.6. The rate of change of the carrier frequency is at the modulating frequency, and the amount of change is proportional to the amplitude of the modulation.

Demodulating FM requires a circuit that is frequency sensitive. The variations in frequency of the incoming signal are converted into variations in voltage at the output of the demodulator. There are a number of circuits that can be used to achieve this – many use a tuned circuit, while others rely on a phase-locked loop. The phase-locked loop variants have the advantage that they normally do not need an inductor. As inductors can be difficult or expensive to produce, and phase-locked loops provide a very high level of performance, their use is often preferred.

Fig 3.5: A single sideband signal consists simply of one sideband. The carrier and one sideband are removed

The degree of frequency variation (deviation) of a frequency-modulated signal is important. It should not pass outside the bandwidth of the receiver. The transmissions used for broadcasting on the VHF bands use wide-band frequency modulation (WBFM) with a deviation of ±75kHz and a bandwidth of 200kHz. Transmissions used for communications purposes have much narrower deviation, typically ±3kHz, and are known as narrow-band frequency modulation (NBFM) signals.

One advantage of FM is that the modulation is carried as variations in frequency. A receiver can be

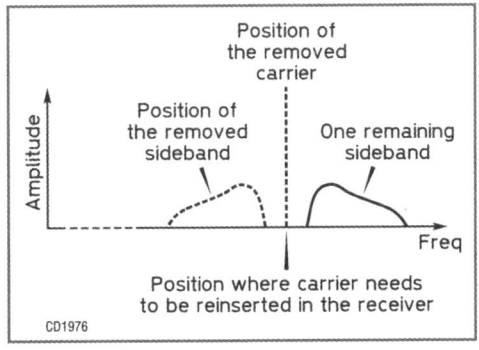

27

# HF AMATEUR RADIO

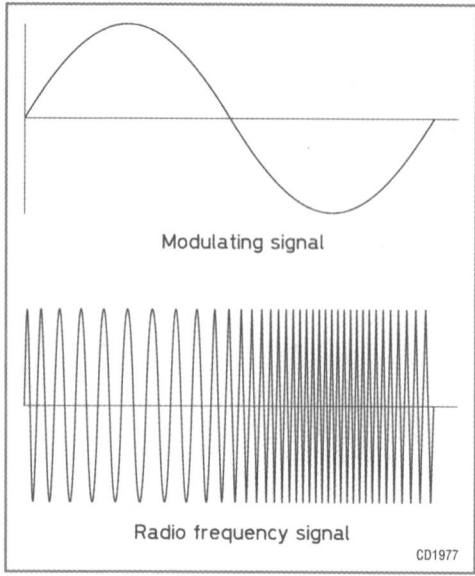

Fig 3.6: Frequency modulation

made sensitive to the frequency variations and insensitive to any amplitude variations. Most noise appears as amplitude variations and so an FM receiver will not be effected by it. The insensitivity to amplitude variations also means that variations in signal level, due to fading for example, will not be as noticeable as with an amplitude-modulated signal. Furthermore, if there is an interfering signal on the channel, the receiver will tend to receive only the strongest signal – a feature known as the capture effect.

Frequency modulation is mainly used for point-to-point communications on the VHF and UHF bands. It is also the preferred mode for mobile and local amateur communications at these frequencies. But it is not widely used on the HF bands except at the top of the 10m band, above 29.5MHz. Having said this, it is possible to DX using FM on the HF band, thanks to a repeater network. It is not unknown for people in their back yard, using a hand-held transceiver, to make world-wide contacts through these repeaters. Simplex operation is generally above 29.6MHz, and repeater outputs are on 29.62, 29.64, 29.66 and 29.68MHz with the input channels 100kHz lower.

## Frequency shift keying

The revolution in digital information has touched most areas of technology and radio, and amateur radio is no exception. It is quite easy to transmit data over a radio link but appropriate forms of modulation need to be used. A system known as frequency shift keying (FSK) is used for many amateur data transmissions. Here, the signal is continuously changed from one frequency to another, one frequency representing the digital one (mark) and the other the digital zero (space). By doing this in step with the digital stream of ones and zeroes, it is possible to send data over the radio.

FSK is widely used on the HF bands. To generate the necessary audio tone from the receiver, a beat frequency oscillator must be used.

On VHF and UHF, a slightly different approach is adopted. An audio tone is used to modulate the carrier and the audio is shifted between the two frequencies. Although the carrier can be amplitude modulated, frequency modulation is virtually standard. Using this audio frequency shift keying (AFSK), the tuning of the receiver becomes less critical.

When the data signal leaves the receiver, it is generally in the form of an audio signal switching between two tones. It needs to be converted into the two digital signal levels, and this is achieved by a unit called a modem (modulator/demodulator). Audio tones fed into its receiver generate the digital levels required for a computer or other equipment to convert into legible text. Conversely, it can convert the out-going digital signals into the audio tones required to modulate the transmitter.

To be able to use these digital modes, several stages of conversion are needed. The audio tones are applied to the modem but some form of decoding is then needed to convert them to a form that can be displayed. And finally a display and keyboard are required. Many stations use a piece of equipment known as a terminal node controller (TNC) to perform the modem function and to undertake processing and transmit-receive switching. A computer or terminal is then used to display the data. However, the most popular method is to use a computer with suitable software and feed the tones into the audio card. In this way, the computer can handle all the functions required without the need for any further equipment.

The speed of the transmission is important. For the receiver to be able to decode the signal, it must know the rate at which the data is arriving. Accordingly, a number of standard speeds are used. These are normally given as a certain number of baud where one baud is equal to one bit per second.

## Radio teletype

The first form of data transmission to gain widespread acceptance was known as radio teletype (RTTY). It was widely used commercially, and it soon became popular with radio amateurs. Large mechanical teleprinters were used to print the data, although nowadays computers are used because they are far quieter and more effective.

Data is sent as a series of pulses, each character consisting of five bits of either a mark or a space. The actual code used is called the Murray code or Baudot code ( Table 3.1). This code is internationally recognised.

The letter shift and figure shift codes are sent to change from upper case to lower case and vice versa. Once case code has been sent, the system will remain in that case until the next case-change code is sent. One of the drawbacks of RTTY is the limited set of characters that can be sent – only text and numbers, and very few other characters.

Data is sent relatively slowly because the mechanical teleprinters could not cope with data any faster. 45.5 baud is the standard amateur speed for HF, although other standards at 50, 56.88, 74.2, and 75 baud exist. The frequency shift between the two tones used to be standardised on 170Hz but now people are increasingly using 200Hz, especially with many of the new digital modes.

As the actual data rate is relatively low, the bandwidths required are less than those needed for SSB. It is easy to gain a rough estimate of the bandwidth needed, by doubling the baud rate and adding the frequency shift. For example, a 50 baud transmission using a 200Hz frequency shift would require a bandwidth of 300Hz, so a 500Hz filter would be quite acceptable.

Given the relatively narrow bandwidths and the need to be able to tune in the tones to about the required frequency, a receiver (or transceiver) with a reasonably slow tuning rate is needed. Typically it should be possible to tune to within about 50Hz.

One of the major problems with RTTY is that any interference causes the received data to be corrupted. Even under relatively good conditions, it is very difficult to have a totally correct copy. When interference levels rise, as they normally do at HF, then copy can be very difficult. To overcome these prob-

# HF AMATEUR RADIO

### Table 3.1: Murray or Baudot Code

| Lower Case | Upper Case | 5(MSB) | 4 | Code Element 3 | 2 | 1 | Decimal Value |
|---|---|---|---|---|---|---|---|
| A | - | 0 | 0 | 0 | 1 | 1 | 3 |
| B | ? | 1 | 1 | 0 | 0 | 1 | 25 |
| C | : | 0 | 1 | 1 | 1 | 0 | 14 |
| D | $ AB | 0 | 1 | 0 | 0 | 1 | 9 |
| E | 3 | 0 | 0 | 0 | 0 | 1 | 1 |
| F | ! % | 0 | 1 | 1 | 0 | 1 | 13 |
| G | & @ | 1 | 1 | 0 | 1 | 0 | 26 |
| H | £ | 1 | 0 | 1 | 0 | 0 | 20 |
| I | 8 | 0 | 0 | 1 | 1 | 0 | 6 |
| J | ' Bell | 0 | 1 | 0 | 1 | 1 | 11 |
| K | ( | 0 | 1 | 1 | 1 | 1 | 15 |
| L | ) | 1 | 0 | 0 | 1 | 0 | 18 |
| M | . | 1 | 1 | 1 | 0 | 0 | 28 |
| N | , | 0 | 1 | 1 | 0 | 0 | 12 |
| O | 9 | 1 | 1 | 0 | 0 | 0 | 24 |
| P | 0 | 1 | 0 | 1 | 1 | 0 | 22 |
| Q | 1 | 1 | 0 | 1 | 1 | 1 | 23 |
| R | 4 | 0 | 1 | 0 | 1 | 0 | 10 |
| S | Bell ! | 0 | 0 | 1 | 0 | 1 | 5 |
| T | 5 | 1 | 0 | 0 | 0 | 0 | 16 |
| U | 7 | 0 | 0 | 1 | 1 | 1 | 7 |
| V | ; = | 1 | 1 | 1 | 1 | 0 | 30 |
| W | 2 | 1 | 0 | 0 | 1 | 1 | 19 |
| X | 1 | 1 | 1 | 1 | 0 | 1 | 29 |
| Z | " + | 1 | 0 | 0 | 0 | 1 | 17 |
| Space | | 0 | 0 | 1 | 0 | 0 | 4 |
| CR | | 0 | 1 | 0 | 0 | 0 | 8 |
| LF | | 0 | 0 | 0 | 1 | 0 | 2 |
| Figure Shift | | 1 | 1 | 0 | 1 | 1 | 27 |
| Letter Shift | | 1 | 1 | 1 | 1 | 1 | 31 |
| Blank | | 0 | 0 | 0 | 0 | 0 | 0 |

AB = Answer back or WHU (Who Are You?)
Upper case characters may vary in some cases as indicated on the chart. Also upper case F, G, and H are not often used.

lems, new data modes have been developed that utilise the power of computers to detect and correct errors.

RTTY is still widely used on the HF bands, and with a data rate of around six characters a second, it is ideal for 'live' chats as most people's typing is not too far off this speed.

## AMTOR

To overcome the problems with RTTY a system known as AMTOR has been developed. It is popular on the HF bands because it gives more reliable communication, especially when interference is present. It achieves this by using a coding system that allows errors to be detected and corrected.

The system uses the same basic five-bit code as RTTY, but sent at a data rate of 100 baud. A total of seven bits are sent. The additional two bits are used to ensure that the transmitted data pattern always contains four mark bits and three space bits. If aware of this expected pattern, the receiver is able to detect an error and action can be taken to correct it.

In operation, the transmitter sends out three characters. The receiver checks them to ensure they are correct. If they are, then an acknowledgement is sent back confirm correct reception. Then the next block of three characters can be sent out. If they have not been received correctly, then this is indicated and the block is re-sent.

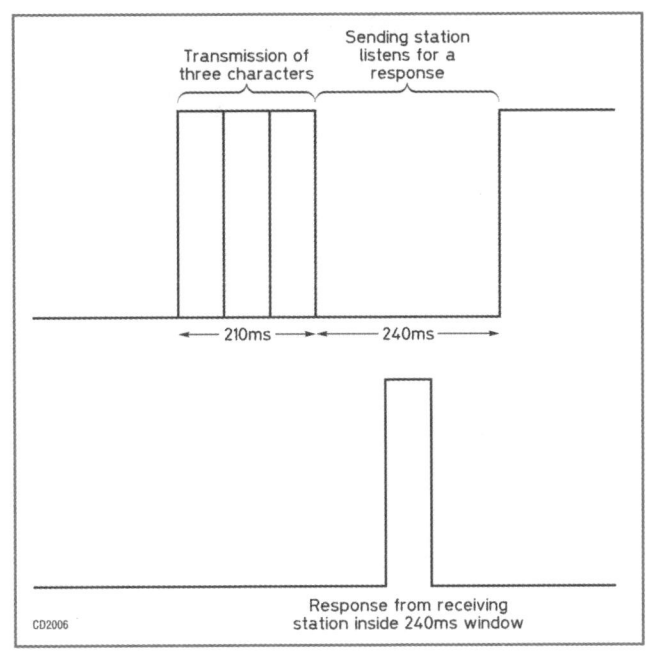

**Fig 3.7: Timing of an AMTOR transmission in Mode A**

A block takes a total of 450 milliseconds (450ms) to send. Each character takes 70ms, giving a total of 210ms for the transmission. Then there is a window of 240ms for an acknowledgement to be received. This amount of time is needed to take account of the delays that occur.

A sending method using an automatic request for repeat (ARQ) is known as Mode A. Mode A transmissions can only operate if contact has been established with a particular station. A general transmission such as a news bulletin or an amateur radio operator wanting a contact cannot work in this mode. Here, a second mode called Mode B is used, where each character is sent twice. Initially the first character is sent once and then the repeat message is sent five characters behind the first one. The time interval between the two signals reduces the possibility of interference causing problems. Sending the data twice also gives the receiver two attempts at capturing each character. Also, because seven bits of data are sent instead of the five used for the character code itself, error detection is still possible, allowing the receiving equipment to decide which character of the two to accept.

Like RTTY, the data rate is around five or six characters a second, depending on the level of interference. And like RTTY, many people use this mode for chatting. It can however also be used for DX. Indeed many people use the mode to look for new countries and rare stations in the same way as they do on Morse and SSB.

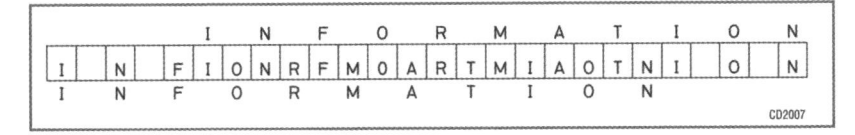

**Fig 3.8: AMTOR Mode B transmission**

## PSK31

This data mode is gaining popularity. It uses phase shift keying rather than frequency shift keying, and it transmits data at a rate of 31 baud. The mode provides greater performance for keyboard-to-keyboard, conversational-style data communications than other data modes.

PSK31 offers an efficient yet straightforward system that does not use the complicated ARQ processes. It provides only enough error correction to match the typical error rates that are encountered. Also, by using phase shift keying and a low data rate, it is possible to narrow the bandwidth, considerably reducing the effects of interference and noise. Bandwidths of 31Hz can be used, making this an extremely narrow-band mode, and one capable of operating under severe conditions.

Phase shift keying is different to the frequency shift keying widely used on the amateur bands. It involves reversing the polarity, or phase, of the signal (180° phase shifts), and has been likened to swapping the two wires in an antenna connection. However, in reality the phase reversals are not achieved in this way; instead they are generated and detected in the audio sections of the SSB transceiver.

This form of modulation is known as binary phase shift keying (BPSK) and it is more efficient than either frequency shift keying, which has a greater bandwidth, or on/off keying, which does not use the power as efficiently.

A novel form of data encoding is used. When sending asynchronous ASCII data, systems use a fixed number of data bits as well as start and stop bits. However, when sending a long run of data it is possible for the receiver to lose synchronisation. Additionally, improvements in speed can be gained from adopting variable-length codes, with codes used most often being the shortest. This technique is also used to good effect in Morse where the character 'e' (which is the most common English letter) is a single dot. By analysing the occurrence of different ASCII characters, a code called Varicode was devised. The shortest code, '00', was allocated to the space between two words.

It is possible to add error correction to the system. However, to achieve this it is necessary to use a form of keying called quadrature phase shift keying (QPSK). Instead of two phase states 180° from one another, QPSK uses four phase states, each 90° from one another. However, QPSK with error correction only sometimes gives better results than ordinary BPSK.

Further information about this interesting mode can be found in reference [1] or by visiting www.psk31.com.

Fig 3.9: A phase reversal in a signal

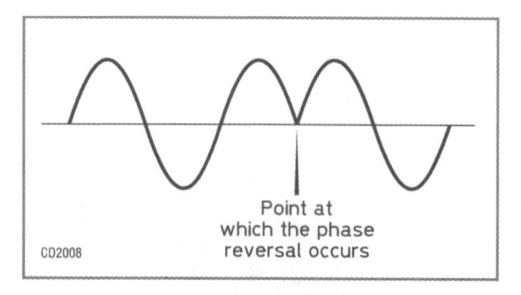

## Slow-scan television

Slow-scan television (SSTV) is used for a number of applications on the HF bands. For example, it is commonly used by radio amateurs to send pictures across the globe, and this adds a completely new dimension to making a contact. Many amateurs transmit pictures of themselves or their station.

SSTV uses many of the basic principles of ordinary broadcast television but such transmissions

CHAPTER 3: TYPES OF TRANSMISSION

**Photo 3.3: A picture transmitted by slow-scan television**

occupy bandwidths of several megahertz, and obviously this cannot be accommodated on the HF bands. SSTV therefore transmits images at a much slower rate so that it can easily be accommodated within the 3kHz bandwidth occupied by a single sideband transmission. Actually, the required transmission bandwidth is between 1.0 and 2.5kHz. However, this bandwidth is sufficient to transmit moving images as well as still pictures.

To create an SSTV signal, a picture is scanned in basically the same way as a normal television signal. A point is moved across the picture and the light level continuously detected. Once the point has completely moved across the picture, it quickly returns to the original side of the picture, but slightly lower, and then it starts again. In this way, the whole of the picture is scanned. At the receiving end the light levels detected at the transmitter are used to build up the original picture. The scanning directions are left to right for the line scan and top to bottom for the frame. Generally there are 120 or 128 lines per frame and the aspect ratio, i.e. the width-to-height ratio, is 1:1.

To ensure that the receiver and transmitter are synchronised, pulses are placed between each line and each complete picture or frame. Line synchronisation pulses are 5 milliseconds (5ms) long, whereas those for synchronising the whole picture or frame are 30ms in length. The picture information has a duration of 60ms.

A typical video signal, as shown in Fig 3.10, is used to modulate the carrier. To achieve modulation, an audio tone is varied in frequency in line with the light level synchronisation pulse. A frequency of 1,200Hz is used for a frame pulse, 1,500Hz for black and up to 2,300Hz for peak white. This is used to give a single sideband signal for transmission. In essence, this generates a radio frequency signal that varies by +400Hz for peak white, -400Hz for black, and -700Hz for a synchronisation pulse. As the synchronisation pulses are represented by a lower frequency than the one representing the black level, they are said to be blacker than black, and they cannot be seen on the screen.

Picture quality can vary widely. To ensure that the optimum quality is achieved, it is essential to achieve good detection of the synchronisation pulses. Often, a special filter is used to detect these 1,200Hz pulses, even though

33

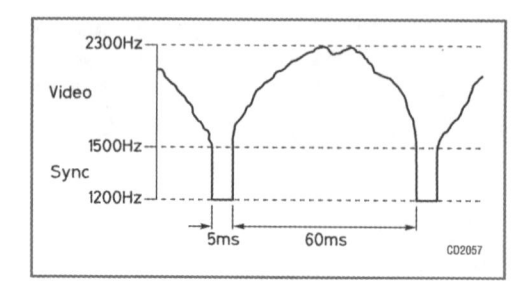

Fig 3.10: A slow-scan television video signal

they can easily be seen in the demodulated video signal.

There are a number of standard picture sizes. Typically, pictures are 128 lines and take eight seconds to send. Another standard is 256 lines, although pictures can be almost any length, terminated by a frame synchronisation pulse.

When slow-scan television first became popular, monitors containing tubes with a long persistence were required to display the pictures. Now most people use PCs running the relevant software to display the images and store them if needed. There are several packages that can be used, making the PC approach far more convenient and cost effective. Using a PC-based system, the audio is extracted from the headphone jack, or a line output socket on the receiver, and applied to a soundcard in the PC.

Computer software is also the most convenient way to generate the pictures. For example, software can create the necessary audio to transmit digital images, whether taken on a digital camera or downloaded from the web, via SSTV.

### Reference:

[1] RSGB Technical Compendium (RadCom 1999), RSGB.

### Further reading:

*Radio Communication Handbook*, 8th edn, ed Mike Dennison G3XDV and Chris Lorek G4HCL, RSGB, 2005.

CHAPTER 4

# Receivers

*In this chapter:*
- Basic receiver concepts
- Direct conversion receiver
- Superhet receiver
- Frequency synthesizers
- Receiver specifications
- Receiver project

RECEIVER performance is of the utmost importance to an HF amateur radio station, particularly when it is being used to pick out weak stations from high levels of interference. Any receiver will be able to receive some stations but leading DXers want something more. They are willing to pay extra to for a receiver that performs very well under exacting conditions. A good receiver, whether a separate item or part of a transceiver, is a key element in any station. However, choosing the right set is not always easy, given the wide range of receivers available. In order to choose a receiver that best meets your needs, it is necessary to understand how the different types of receiver work. This knowledge will also help you get the best performance out of a receiver.

Photo 4.1: A typical HF receiver

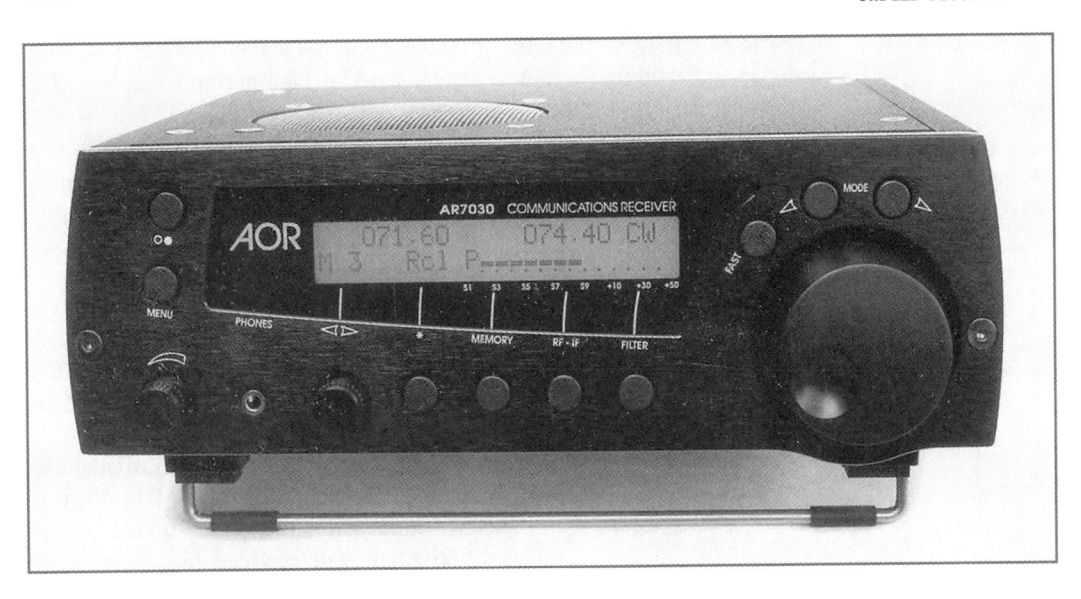

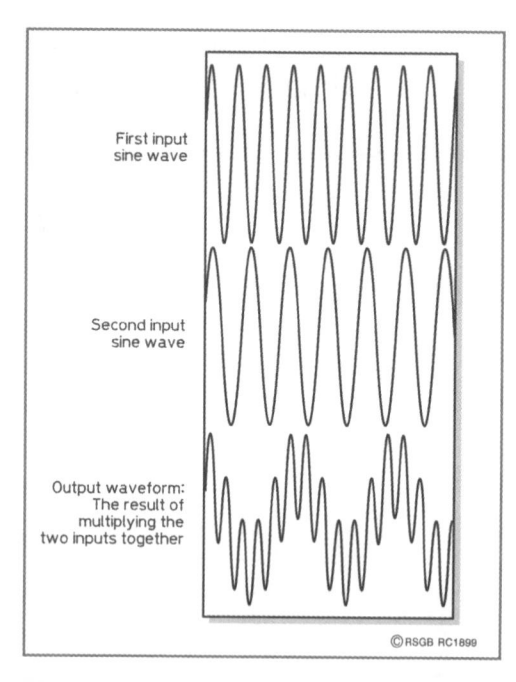

Fig 4.1: The output from a mixer as seen on an oscilloscope

Fig 4.2: Signals produced by a mixer

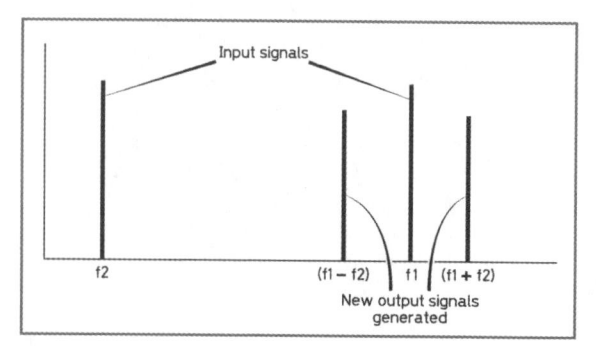

## Receiver types

There are many HF receivers available on the market today. They largely fall into two major categories. The first is the direct-conversion receiver where signals are mixed with an oscillator to give an audio output directly. The second main type is the superhet, where the in-coming signal is mixed with a local oscillator to produce an output at an intermediate frequency. This is filtered and then the signal is demodulated to give the required audio. Both of these receivers rely on the principle of mixing for their operation.

## Mixing

The mixing is not an additive process like that in an audio mixer. Instead, it is a multiplication process where the output from the mixer is the product of the two inputs. A typical waveform is shown in Fig 4.1. The output contains signals with two new signals at the sum and difference frequencies of the input signals. If the input signals are at frequencies of f1 and f2, then new signals are produced at frequencies of (f1 + f2) and (f1 - f2). For example, if the input frequencies are 1MHz and 100kHz, then new signals will be generated at 900kHz (difference frequency) and 1.1MHz (sum frequency).

## Direct-conversion receiver

This type of receiver can be relatively simple but it can still provide an excellent level of performance. As a result, it is very popular, especially with low-power enthusiasts who build much of their own equipment. However, it does have some limitations. It doesn't offer the flexibility and facilities of a superhet receiver, and therefore is not generally used in high-performance sets. The basic block diagram for this type of receiver is shown in Fig 4.3.

The signals from the antenna enter the mixer along with a signal from a variable-frequency oscillator. The signals are mixed together, and the signals from the antenna that are close in frequency to that of the local oscillator produce audible 'beat note' signals. These pass through the audio low-pass filter and can be amplified to be heard in a loudspeaker. Those signals that are not close in frequency will produce mix signals that fall above the audio band and will not pass beyond the audio frequency low-pass filter.

As an example, if the local oscillator is set to a frequency of 2.000MHz, then an incoming signal on a frequency of 2.001MHz will produce mix products at

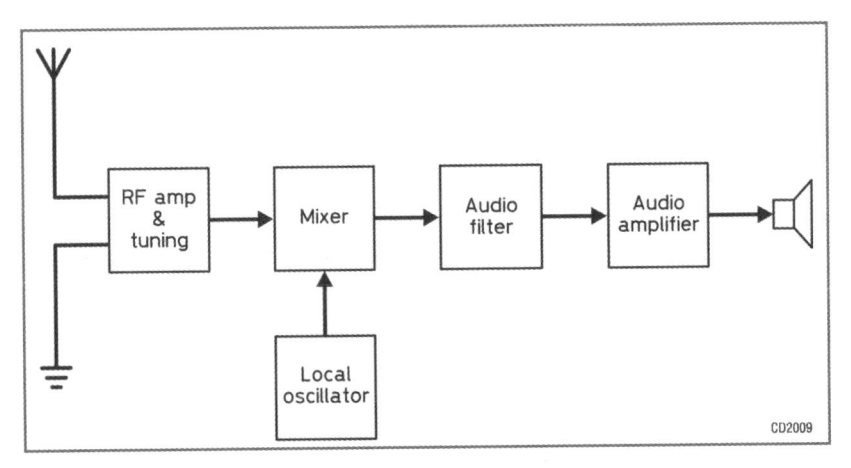

Fig 4.3: Block diagram of a direct conversion receiver

0.001MHz (i.e. 1kHz) and 4.001MHz. The signal at 4.001MHz is well above the pass band of an audio low-pass filter and is therefore removed. The one at 1kHz is within the audio bandwidth and will pass through the filter to be amplified and fed into the loudspeaker. Another incoming signal on the slightly higher frequency of 2.040MHz will produce signals at the output of the mixer at frequencies of 4.040 and 0.040MHz. Both of these will be above the cut-off point of the low-pass filter and will not be heard. For communications purposes, the cut-off frequency of the audio filter would be set to around 3kHz. In this way, signals up to 3kHz either side of the local oscillator will be heard. Signals on different frequencies can be heard by varying the frequency of the local oscillator.

Some radio frequency tuning is normally undertaken prior to the mixer. This limits the signals entering the mixer to those in the region of the frequency of interest. Although the tuning does not have to be very sharp, it helps prevent the mixer from becoming overloaded by having signals from the whole radio spectrum appearing at its input.

The major disadvantage of direct-conversion receivers is that they operate by creating a beat note with the incoming signal that appears within the audio bandwidth. This is exactly what is required for Morse (CW) and SSB, but it is not ideal for AM where the receiver has to be tuned to be zero beat with the carrier. It is also not possible to resolve FM.

A further disadvantage is what is known as the audio image. This occurs because signals either side of the local oscillator appear within the audio band and can be heard. Using the example figures above, signals at both 2.001 and 1.999MHz will produce beat note signals of 1kHz. This means that levels of interference are higher than for an equivalent superhet where this problem is not experienced.

## Superhet receiver

The superhet is the most widely used form of receiver. Invented by Edwin Armstrong in 1918, it provides the flexibility and performance required by most amateurs. It operates by mixing the incoming signals with a locally generated oscillator signal to convert them down to a fixed intermediate frequency

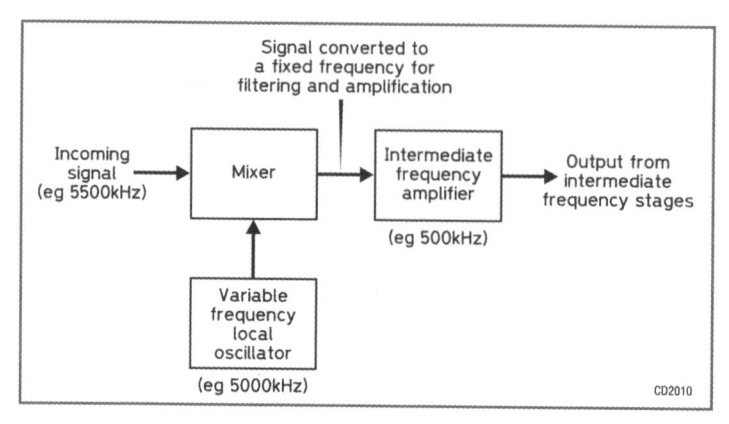

# HF AMATEUR RADIO

**Fig 4.4: The principle of the superhet receiver**

**Fig 4.5: Front-end tuning to remove the image signal**

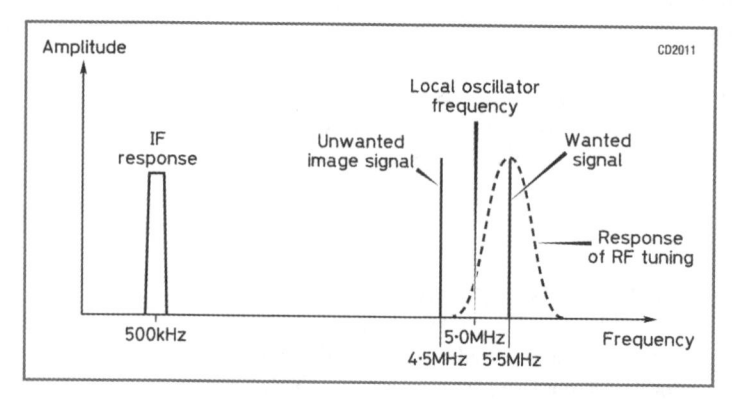

(IF). It is within the IF stages of a receiver that the selectivity is provided which enables the unwanted off-channel signals to be rejected. By changing the frequency of the local oscillator, signals on different frequencies can be converted down to the IF and hence the receiver can be tuned.

One problem with the system is that it is possible for signals on two different frequencies to enter the IF. Take the example of a receiver with an IF of 500kHz and a local oscillator tuned to 5MHz. A signal with a frequency of 5.5MHz will mix with the local oscillator to give an output at 500kHz. This will pass through the IF filters. It is also possible for a signal on 4.5MHz to mix with the local oscillator to give an output at 500kHz. To prevent this happening, a tuning circuit is required before the mixer so that only the required frequency band is allowed through as shown in Fig 4.5. The unwanted signal is known as the image.

Looking at the overall block diagram of the receiver it can be seen that the signal first enters the RF amplifier stage. The purpose of this stage is two-fold: to provide sufficient selectivity to prevent image signals reaching the IF stages, and to provide signal amplification before the mixer.

From the RF amplifier, the signal passes to the mixer where it is converted down to the intermediate frequency. While the block diagram only shows one conversion, this process may be undertaken in several stages. The reason for this is that if a high intermediate frequency is chosen, the image frequency will be further away from the wanted one, enabling the image response to be improved. A high IF will also mean that achieving the required level of adjacent channel selectivity is more difficult, whereas a low IF will mean the image rejection is poor. Many receivers adopt a two- or even a three-stage conversion process to provide the required balance between all the conflicting factors.

Often HF receivers may actually convert the incoming signal up in frequency at the first conversion to achieve the required image rejection. Afterwards, it is converted down in frequency to enable the required degree of adjacent channel selectivity to be achieved.

The variable-frequency local oscillator is important, and a variety of approaches

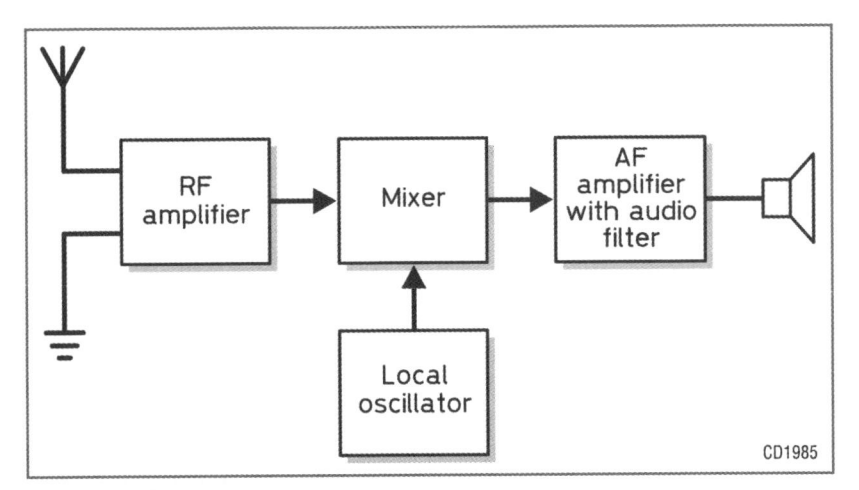

Fig 4.6: Block diagram of a basic superhet receiver using a single conversion. Some sets have two or three conversions to give the required level of performance as described in the text

can be used. The most straightforward is to employ a free-running, LC-tuned oscillator (i.e. an oscillator using a capacitor and inductor for the resonant circuit that determines the frequency of its operation). These were widely used up until the 1970s but had the drawback that they tended to drift in frequency. Nowadays, frequency synthesizers are used. These circuits offer many advantages in terms of stability and flexibility of operation. They are described later.

The IF circuits in a receiver provide the majority of amplification and selectivity. In receivers using discrete components, several stages of amplification are used. Nowadays, however, just one integrated circuit is likely to be employed to give the required level of gain. As the IF provides the adjacent-channel selectivity for the receiver, the filter used in these stages governs the performance of the whole set. Accordingly, a high-performance filter is used, which in most sets is a crystal type.

The output from the IF amplifier is connected to the demodulator. As different types of demodulator are needed for different types of transmission, there may be two or more different circuits that can be switched in. For single sideband or CW, product detectors or demodulators with a beat frequency oscillator (BFO) are used. Essentially, the product detector is a mixer and the BFO beats with the incoming signal to give an audio note in the case of CW or regenerates the audio signals in the case of single sideband.

Once the signal has been demodulated, it can be amplified and connected to a loudspeaker, headphone or, in the case of data transmissions, it can be connected to the data equipment.

## Frequency synthesizers

Frequency synthesizers are used almost universally for local oscillators in receivers and transceivers today. There are two forms of synthesizer: those based on phased locked loops that are often called indirect synthesizers, and those which synthesise a signal using only digital techniques. This latter are known as Direct Digital Synthesizers (DDS). Many sets that include a DDS use a phase locked loop as well, thereby combining both technologies in the same synthesizer.

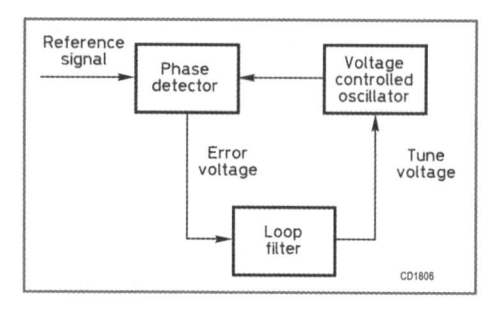

**Fig 4.7: A basic phase-locked loop**

Both these synthesizers offer many advantages over other techniques. For example, they can be very easily controlled by microprocessors, their frequency stability is excellent and they are exceedingly versatile, allowing many facilities to be incorporated into sets, including scanning, multiple VFOs and memories.

The phase-locked loop (PLL) is the most popular of the two. A basic loop is shown in Fig 4.7. It consists of a phase detector, voltage-controlled oscillator (VCO), a loop filter and finally a reference oscillator. As the frequency stability of the loop or synthesizer is totally governed by the reference oscillator, this is crystal controlled, and in professional equipment it often uses a temperature-controlled oven.

The operation of the basic phase-locked loop is fairly straightforward. The reference oscillator and VCO produce signals that enter the phase detector. Here, an error signal is produced as a result of the phase difference between the two signals. This signal or voltage is then passed into a filter that serves several functions. It controls the loop stability, defines many of the loop characteristics and also reduces the effect of any sidebands that might be caused by reference signals appearing at the VCO input.

Once through the filter, the error voltage is applied to the control input of the VCO so that the phase difference between the VCO and reference is reduced. When the loop has settled and is locked, the error voltages will be steady and proportional to the phase difference between the reference and VCO signals. As the phase between the two signals is not changing, the frequency of the VCO is exactly the same as the reference.

In order to use a phase-locked loop as a synthesizer, a divider is placed in the loop between the VCO and phase detector. This has the effect of raising the VCO signal in proportion to the division ratio. Take the example when the divider is set to 2. The loop will reduce the phase difference between the two signals at the input of the phase detector, i.e. the frequency of both at the two phase detector inputs will be the same. The only way this can be true is if the VCO runs at twice the reference frequency. Similarly, if the division ratio is 3, then the VCO will run at three times the reference frequency. In fact, by mak-

**Fig 4.8: A basic frequency synthesizer loop**

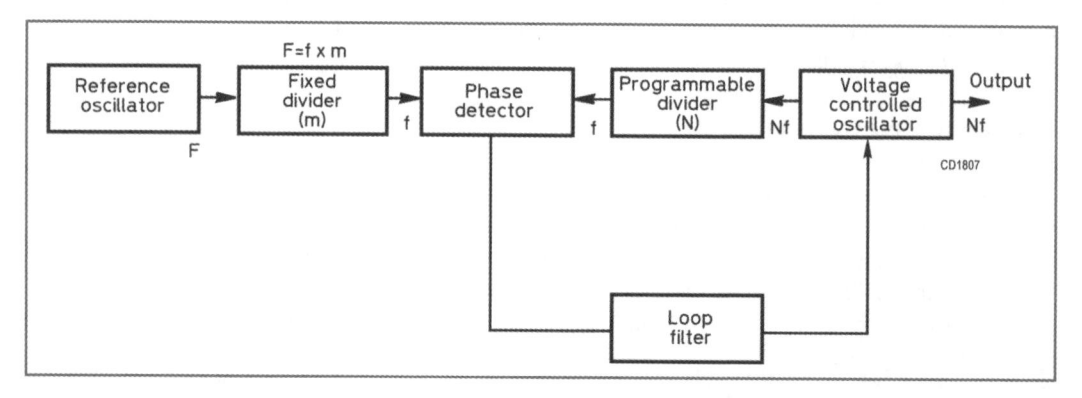

ing the divider programmable, the output frequency can be easily changed.

In order to have channel spacing of 25kHz or less, the reference frequency has to be made very low. This is generally done by running the reference oscillator at a relatively high frequency, eg 1MHz, and then dividing it down as shown in Fig 4.8.

This is the basic synthesizer loop that is at the heart of virtually all frequency synthesizers. It can be enhanced in many ways to give more flexible operation and smaller step sizes that give almost continuous tuning.

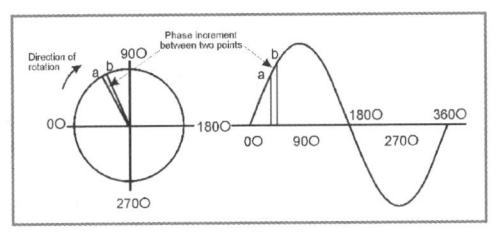

Fig. 4.9: Operation of the phase accumulator in a direct digital synthesizer

The second form of synthesizer that is widely used is the DDS. As the name suggests, this method generates the waveform directly from digital information. The synthesizer operates by storing various points in the waveform in digital form and then recalling them to generate the waveform. In any system, phase advances around a circle as shown in Fig. 4.9. As the phase advances around the circle, this corresponds to advances in the waveform as shown. The idea of advancing phase is crucial to the operation of the synthesizer as one of the circuit blocks is called a phase accumulator. This is basically a form of counter. When it is clocked, it adds a preset number to the one already held. When it fills up, it resets and starts counting from zero again. In other words, this corresponds to reaching one complete circle on the phase diagram and restarting again.

Once the phase has been determined it is necessary to convert this into a digital representation of the waveform. This is accomplished using a waveform map. This is a memory which stores a number corresponding to the voltage required for each value of phase on the waveform. In the case of a synthesizer of this nature, it is a sine look up table. In most cases, the memory is either a read only memory (ROM) or programmable read only memory (PROM).

The next stage in the process is to convert the digital numbers into an analogue voltage using a digital to analogue converter (DAC). This signal is filtered to remove any unwanted signals and amplified.

## Selectivity

The selectivity of a receiver governs the way it accepts signals on the wanted frequency and rejects those that are off-channel. The most obvious way this affects a receiver is in terms of its adjacent channel selectivity.

Fig. 4.10: Block diagram of a direct digital synthesizer

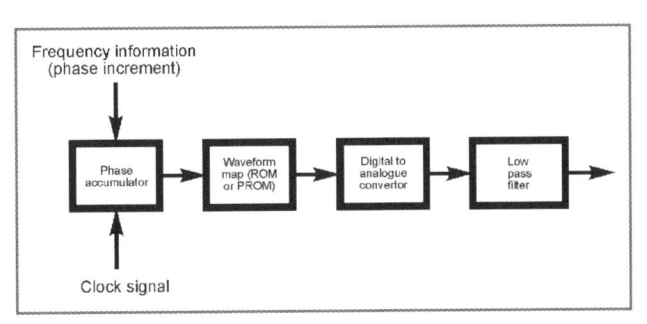

The adjacent-channel selectivity is governed by the performance of the IF filter. Its performance will determine how well the whole receiver performs in rejecting signals that are just off-channel. It is worth noting that for different types of emission, different bandwidths are required. For Morse transmissions, bandwidths of 250 or 500Hz are mostly used, while for

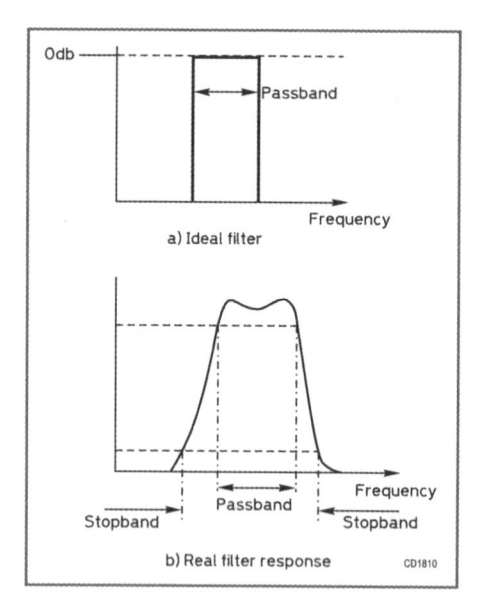

**Fig 4.11: Response of ideal and real filters**

SSB, figures of between 2.5 and 3kHz are common.

In an ideal world, the response of a filter would be as shown in Fig 4.11(a). However, in reality, it will look more like Fig 4.11(b). By comparing these, it can be seen that there is a significant difference.

The two main specifications are the bandwidths of the pass band and the stop band. The bandwidth of the pass band is the bandwidth between the two points where the voltage response drops by 6dB (i.e. to a quarter of the power level) from its in-band level. This is the figure quoted as the bandwidth of the filter above and is taken as the range of frequencies that are accepted by the filter. The stop band is equally important. It is normally the bandwidth where the response has fallen to a level where it should not let signals through. This is normally taken as the point where the response has fallen by 60dB from its in-band level.

The difference in bandwidth between the pass band and stop band shows how quickly the response falls away. Ideally, they should be almost the same but this is never the case in reality. To give an indication of the shape of the filter response, a figure known as the shape factor is sometimes quoted. This is simply the ratio between the pass band and the stop band bandwidths. Thus a filter having a bandwidth of 3kHz at 6dB and 6kHz at 60dB would have a shape factor of 2:1. For this figure to have real meaning, the two attenuation figures must also be quoted. For example, a specification might quote a shape factor of 2:1 at 6/60dB. There are several different types of filter that can be used in a receiver. The simplest types are LC filters, and in many older sets the interstage coupling transformers are tuned to give the selectivity. A typical circuit showing how this is accomplished is shown in Fig 4.12. Normally, one transformer is found on the output of the mixer, one for each interstage coupling that is required and one on the output of the IF strip prior to the demodulator. In broadcast sets, for example, three transformers may be found. The problem with these LC filters is that they do not give the very high levels of selectivity required on today's amateur bands.

Crystal filters are normally used to achieve the very high selectivity needed by amateur receivers and transceivers. These use the piezo-electric effect for their operation. On the one hand, an electrical current causes sympathetic mechanical vibrations in the crystal. Similarly, mechanical vibrations on the crystal will give rise to an equivalent electrical signal. In this way, the electrical circuit may use the mechanical resonances of the crystal, and as these possess very high degrees of selectivity or Q, they are able to be used to produce very-high-performance filters. Normally, several crystals are used to achieve the required performance.

In some cases, filter specifications for a receiver or transceiver may include a figure for the number of poles in the filter. To explain this in any detail requires looking at some filter theory. However, it is sufficient to say that there is a pole for every crystal in the filter. For example, an eight-pole filter has eight crystals. Most filters today have six or eight poles.

While most of the filtering is performed in the IF stages of a superhet receiver, sometimes filtering in the audio stages can be used to very good effect. Normally, audio filters are designed around operational amplifiers using only capacitors and resistors. With these circuits, very high degrees of selectivity can be achieved without the use of expensive components. Often these filters are used in the reception of Morse signals where very narrow bandwidths can be used. The main disadvantage of an audio filter is that the AGC is derived from the signal appearing at the output of the IF. This can be a disadvantage if there is a strong signal within the IF pass band but outside the pass band of the audio filter. In such cases, the strong signal will 'capture' the automatic gain control (AGC), altering the gain level of the receiver. A weak signal within the pass band of the audio filter will then be altered in strength by the AGC level variations determined by the strong signal.

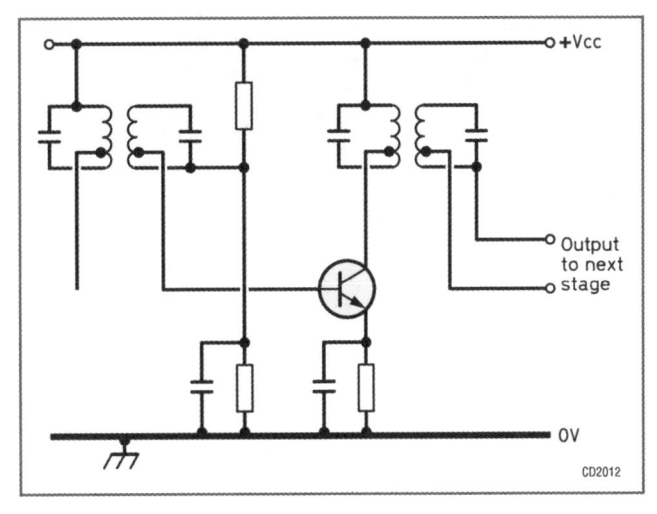

**Fig 4.12: A typical IF stage using a transformer and discrete components**

## Image response

The image response of a receiver is very important. If the performance of the receiver is poor in this respect, unwanted signals will appear and cause interference, possibly masking the wanted signals. This can be a particular problem if weak signals are being received on a quiet band, and the image frequency falls on a band where there are strong signals, for example from broadcast stations using many hundreds of kilo-watts.

Image response rejection figures are normally expressed as a certain number of decibels rejection at a given frequency. For example, the image rejection may be 60dB at 30MHz. This means that a signal on the image frequency must be 60dB greater in strength than one on the wanted channel to produce an output of the same level. Although this may appear to be a large amount, signals picked up by receivers vary in strength by enormous amounts and often more than 60dB.

Image rejection varies with the frequency of the receiver, degrading at the higher frequencies. The reason for this is that the difference between the wanted signal and the image is a smaller proportion of the operating frequency.

## Noise performance and sensitivity

The noise performance of a receiver is very important. If a receiver introduces any noise onto the signal, then it will tend to mask it out, reducing readability or in extreme case completely obliterating it. The noise performance of the first stages in the receiver are the most important. As each successive stage within the receiver will amplify noise from the previous stages, this means that the

# HF AMATEUR RADIO

**Fig 4.13: Noise field strength values in a 2.7 kHz bandwidth**

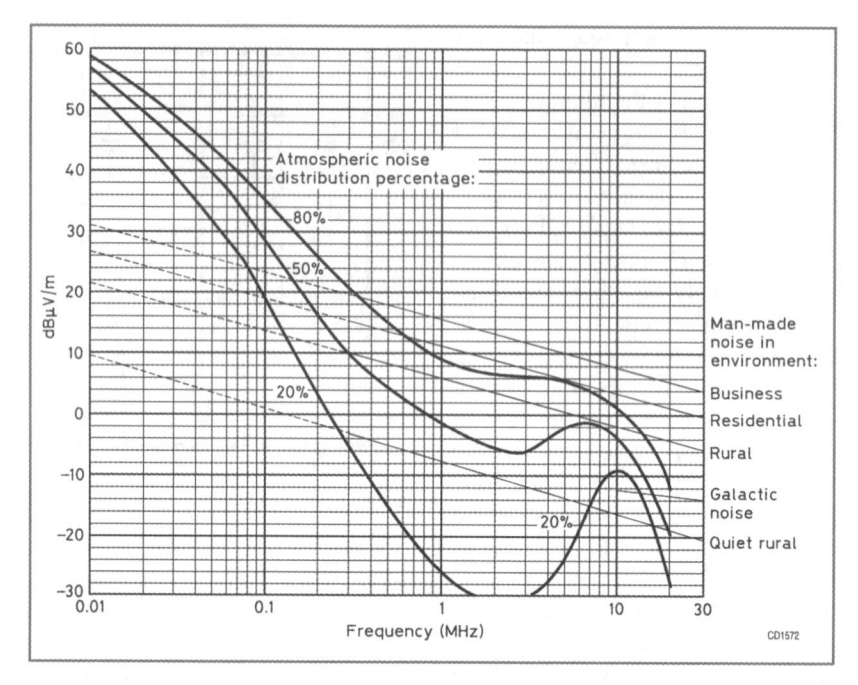

design of the first RF amplifier is critical and should be optimised to give the lowest noise.

While the noise performance of the receiver is important, it is not as crucial as it is for sets used on the VHF and UHF bands. The reason for this is that the level of noise picked up by the antenna on the HF bands generally exceeds that generated in the receiver. This noise comes from a variety of sources (Fig 4.13). Some is man made, and this is obviously lower in rural areas, and tends to be highest in business or industrial areas. Another constituent of the received noise is atmospheric noise. The level of this is dependent on a variety of factors, including the time of day and atmospheric conditions. Some galactic noise may also be present.

The simplest method of quantifying the noise performance of a receiver is to quote the signal-to-noise ratio for a given input signal. The measurement of the difference between the signal and the noise indicates the noise performance and hence the sensitivity of the receiver (Fig 4.14).

**Fig 4.14: Concept of signal to noise ratio**

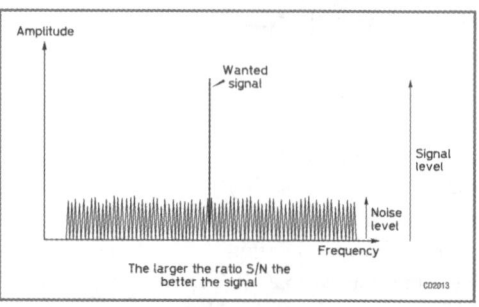

For this measurement to be meaningful, the bandwidth of the receiver must also be given. The reason for this is that the wider the bandwidth, the greater the level of noise that is received. When used with amplitude-modulated signals the depth of modulation must also be given because this will vary the level of the output signal from the loudspeaker. This method is widely used for HF receivers, and here a figure of around 1 microvolt for a 10dB signal-to-noise ratio in a 3kHz bandwidth for SSB or Morse reception is common. For AM, a figure of 1.5 microvolts for a 10dB signal-to-noise ratio in a 6kHz bandwidth at 30% modulation may be achieved.

# CHAPTER 4: RECEIVERS

## Strong-signal handling performance

Although it is important for any receiver to be sensitive, it is equally important for it to be able to withstand strong signals without overloading. If it cannot handle strong signals well, then the more interesting ones that are often quite weak may be masked out by the effects of overloading.

There are a number of parameters that effect the strong-signal handling capability of a receiver. Two of the most important are the third-order intercept and the third-order intermodulation distortion specifications.

Receiver amplifiers do not have an infinite signal-level range. The output will follow a linear relationship for input signal levels up to a certain point. Beyond this point, the amplifier is unable to provide the required output level and the signal becomes compressed. Unfortunately, the non-linearity brings a number of problems with it, including intermodulation products, cross-modulation, and blocking.

A number of problems arise when an amplifier runs into non-linearity. Harmonics are generated and signals mix with one another. These products on their own do not normally cause a problem as the front-end selectivity means that signals that could enter the amplifier generate signals that are outside the range of interest. Take the example of a receiver set to around 10MHz. If there is a strong signal at 10.1MHz, then the harmonics will fall at 20.2, 30.3MHz and so forth. These are well outside the range of the receiver and they are clearly going to have no effect on reception. Similarly, signals that could enter the amplifier and mix with one another also fall outside the range of interest. Take the example of signals at 10.0MHz and 10.1MHz. They will mix to produce signals at 0.1 and 20.1MHz, and these are clearly not going to affect reception.

However, problems arise when a harmonic of one signal mixes with the fundamental of another, e.g. (2f1 f2), this being known as a third-order product. The third harmonic of one may mix with the second harmonic of the other, e.g. (3f1 2f2), a fifth-order product. To put figures against the formulae, take two signals at 10.0 and 10.1MHz. The third-order product is 9.9MHz, and the fifth-order product is 9.8MHz.

Similarly, signals are produced at 10.2 and 10.3MHz. In fact, a 'comb' of signals stretches out from the two main signals, each separated by the frequency difference between them (see Fig 4.16). As these signals are in the same region as the incoming signals, it is possible for them to pass into later stages of the receiver.

## Third-order intercept

To gain an idea of the resilience of a receiver to intermodulation distortion, a concept known as the third-order intercept point is used. The figure uses the fact that the third-order products rise in level much faster than the wanted ones as the input level rises.

Fig 4.15: Compression in an amplifier results in the output being compressed beyond a certain in-put level, giving rise to problems such as intermodulation, cross-modulation and blocking

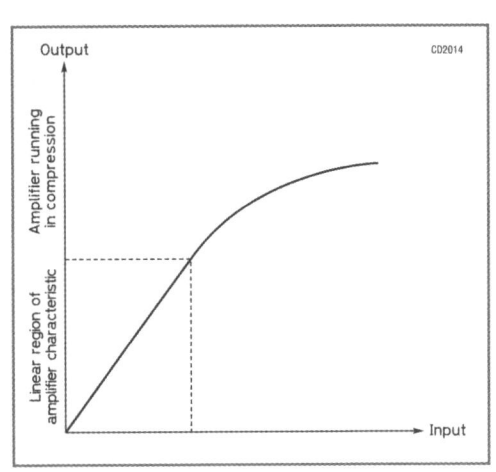

# HF AMATEUR RADIO

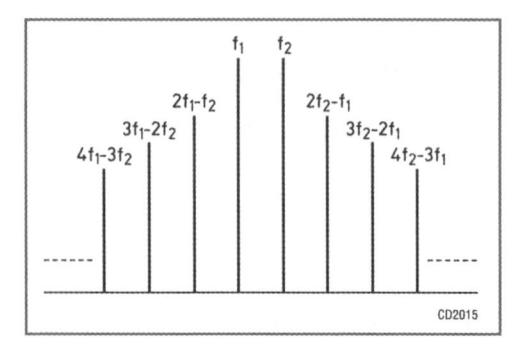

Fig 4.16: Intermodulation products stretch out either side of the two incoming signals. It is possible for two strong signals to pass through the RF tuning, causing the RF amplifier to run into compression and generating intermodulation products that can enter the following stages

A plot can be made of the output for varying input levels. As the input signal level is increased, the output level rises in line with it. Initially, no third-order products are seen because they are masked out by the noise but, as the input level increases, they start to rise. As the input level is increased still further, the amplifier will start to 'limit' and, if the curves for the levels of the wanted signals and the third-order products are extrapolated, it will seen that they cross. The point at which this occurs is known as the third-order intercept point and it is measured in watts – or in most cases milliwatts. The higher the level of the intercept point, the better the strong-signal handling of the receiver. Typically, an amateur set might have an intercept point of 15dBm, i.e. 30 milliwatts (30mW).

## Blocking

Sometimes when a very strong signal is received, it can 'block' the front-end of the receiver, reducing the level of other signals. This is a problem if it happens when a weak signal is being received, and can arise during contests when many very strong signals may be present on the band.

Blocking occurs because the receiver front-end runs into compression. When this happens, it has the effect of reducing the level of the other signals passing through it in a similar way to the capture effect encountered in the reception of FM signals.

The amount of blocking is dependent on the level of the incoming signal. Another factor is the frequency offset from the channel being monitored – the further away the strong signal, the less is its effect because the selectivity, particularly in the front-end stages, reduce its level. Blocking is specified as the level of the unwanted or strong signal which reduces the gain of the receiver by 3dB at a given offset which is normally 20kHz. A good receiver should be able to withstand signals of more than about 10mW before the gain is reduced by 3dB.

## Cross-modulation

This is another effect that can be noticed on occasions, and was particularly important when amplitude-modulated signals were more popular. When it occurs, the amplitude variations or modulation on a nearby strong signal are heard on weaker signals close by.

Cross-modulation is a third-order effect, and a receiver with a good third-order intercept point is likely to have good cross-modulation performance.

To specify cross-modulation performance, the effect of a strong amplitude-modulated carrier with a known level of modulation is noted on a small wanted signal. The specification involves several figures. The strength of the strong signal must be stated, along with the level of modulation, the frequency offset from the weak wanted signal and the level of the wanted signal. Generally, the specification is worded as the level of a strong carrier with 30% modulation to produce an output 20dB below that produced by the wanted signal with a 20kHz offset. Usually, the wanted signal level is taken as 53dBm (i.e. 53dB below 1mW), and into a 50-ohm load this works out to be 1mV.

## Dynamic range

Both sensitivity and strong-signal handling capacity are important in any receiver. Accordingly, the dynamic range, or range over which the receiver can operate, is very important. It is defined as the difference between the weakest signal that the receiver can receive and the strongest one it can tolerate without any noticeable degradation in performance. Unfortunately, there are a number of ways in which the points at both ends of the scale can be measured and care needs to be taken when assessing dynamic range specifications.

The weakest signal that can be received is governed by the sensitivity. A term called the minimum discernible signal (MDS) is often used. It is generally taken as a signal equal in strength to the noise produced by the set, and its level is generally given in dBm, i.e. decibels relative to a milliwatt for a given receiver bandwidth. The bandwidth must be included because the level of noise is dependent on this. Typically it might be around 135dBm for a bandwidth of 3kHz.

At the other end of the scale, there are two main factors that come into play. One is the generation of intermodulation products, and the other is blocking. As the onset of these effects occurs at different signal levels, the way in which the high end of the dynamic range is determined must be mentioned in the specification.

Even when blocking is used to determine the high end of the specification, different manufacturers use different levels of blocking. Often a 1dB decrease in sensitivity is used, but sometimes 3dB is used instead.

Where the level of intermodulation products is used to determine the top end of the range, it is normally taken that the level of these products must not exceed the MDS, ie be no greater than the noise floor of the set.

Most modern sets have an intermodulation-limited dynamic range of between 80 and 95dB. If a specification using the blocking figure is taken instead, then a range of 115dB or possibly even more might be expected. The difference between figures obtained by the different methods highlights the care that must be taken when comparing sets from different manufacturers where different measurement methods may be used.

## Phase noise and reciprocal mixing

Frequency synthesizers have many advantages but one of their main disadvantages is that some designs generate large amounts of phase noise. Some such noise is present on any oscillator signal, and essentially it manifests itself as a small amount of frequency variation or 'jitter' on the signal. If the signal is examined on a spectrum analyser, it is seen as noise spreading out either side of the wanted signal as shown in Fig 4.17.

Different types of oscillator produce different levels of phase noise. Crystal oscillators are generally the best, and LC-tuned variable frequency oscillators are normally very good. In general, the higher the Q (quality factor) of the

# HF AMATEUR RADIO

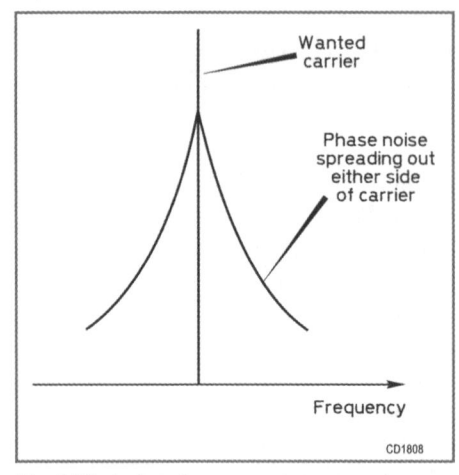

**Fig 4.17: A signal with phase noise**

(Figure shows wanted carrier with phase noise spreading out either side of carrier, plotted against frequency.)

**Fig 4.18: Reciprocal mixing**

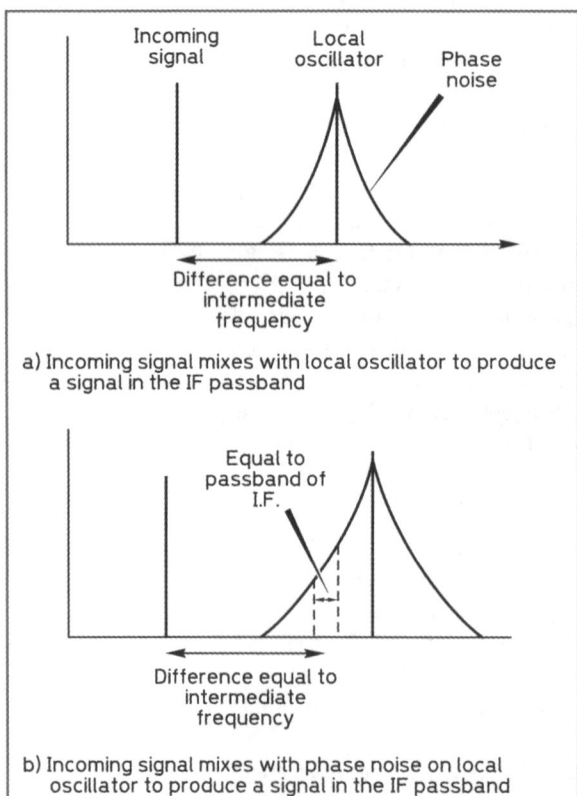

a) Incoming signal mixes with local oscillator to produce a signal in the IF passband

b) Incoming signal mixes with phase noise on local oscillator to produce a signal in the IF passband

tuned circuit, the lower the level of phase noise. However, the way in which frequency synthesizers operate means that some can produce high levels of phase noise.

The main way in which phase noise affects receiver performance is through what is termed reciprocal mixing. To look at how this occurs, take the example of a superhet receiver tuned to a strong signal. The signal will pass through the radio frequency stages to the mixer where it will be converted down to the intermediate frequency stages, passing through the filters to the demodulator. When the receiver is tuned off frequency by 10kHz, for example, the signal will no longer be able to pass through the intermediate frequency filters. However, it will still be possible for phase noise on the local oscillator signal to mix with the incoming signal to produce noise that will fall inside the receiver pass band as shown in Fig 4.18. In some cases, this might be strong enough to mask out a weak wanted signal.

A number of different methods are used to define the level of reciprocal mixing. Generally, they involve measuring the response of the receiver to a large off-channel signal. These measurements are rarely easy as the performance of any signal generator must be very good – significantly better than the receiver – otherwise the performance of the signal generator is measured.

A measurement can be made by noting the level of audio from a small signal when the BFO is switched on. The signal is then tuned off-channel by a given amount, often around 20kHz, and the signal increased until the same audio level is achieved as a result of the receiver phase noise. As the level of noise depends on the receiver band-width, this has to be stated in the specification as well. A good HF receiver might have a figure of 95dB at a 20kHz offset using a 2.7kHz bandwidth. This figure will naturally improve as the frequency offset is increased. At around 100kHz one might expect to see a figure of more than 105dB.

Another way of measuring the phase noise is to inject a large signal into the receiver and monitor the level needed to give a 3dB increase in background noise. As a variety of measurements can be used,

# CHAPTER 4: RECEIVERS

it is often best to study magazine reviews written by the same person because he will use consistent measurement techniques. In this way, several sets can be compared against one another.

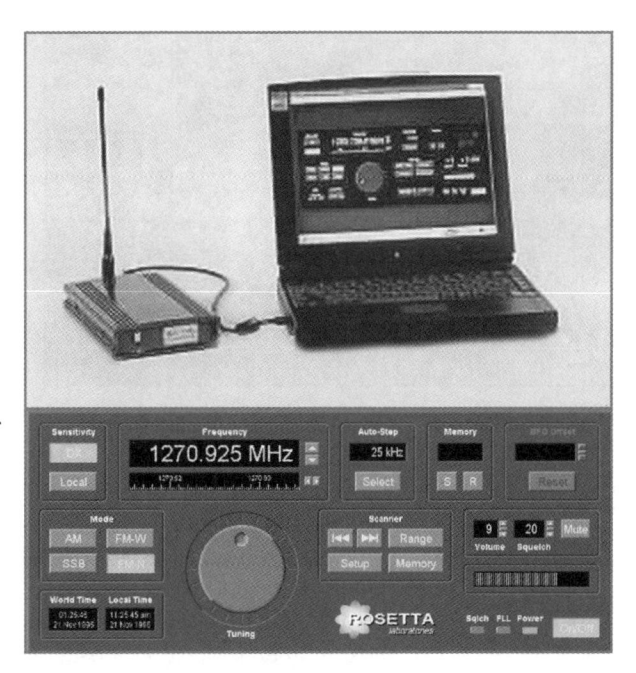

## Digital signal processing

Today many new pieces of equipment advertise the fact that they use digital signal processing (DSP) techniques. As the technology for DSP becomes cheaper, it will be used increasingly in the years to come, giving the added advantage of improved performance.

DSP involves converting a signal into a digital format and then processing the signal using computer techniques. This processing is able to perform the actions that have traditionally been performed by analogue components like filters, amplifiers, mixers and demodulators. As these functions are performed mathematically, they do not have many of the limitations of the analogue components, and they are able to achieve much higher performance. While this may seem like a dream come true, there are still limitations and it is not, for example, possible to produce the perfect filter. Another advantage is that further functionality or improved filters can be added simply by changing the software.

The first step in the process is to change the analogue signal into a digital format. This is achieved using an analogue-to-digital converter (AD converter). The signal is sampled at regular intervals of time, and the voltage at that time is converted into a digital format as shown in Fig 4.19.

Once the signal is in digital format, it is passed into the signal processing section. The core of this section is generally a specially designed processor, optimised for digital signal processing. Here, the sequential values representing the signal are processed mathematically to provide in a digital way all the functions that would normally be provided by analogue components.

Once the processing has been undertaken, the signal is often converted back into an analogue format using a digital-to-analogue converter (DA converter) to be passed into an audio amplifier and then to headphones or a loudspeaker as required.

With the advances being made in technology, DSP is becoming cheap-

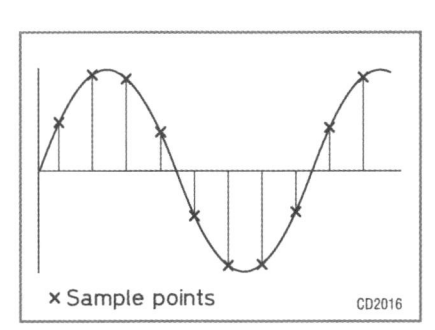

**Photo 4.2:** Winradio uses the processing power of a PC to perform the digital signal processing, control of functions like the tuning as well as providing the interface for the receiver controls (photos courtesy Winradio)

**Fig 4.19: Converting a signal into a digital format by sampling at regular intervals of time**

# HF AMATEUR RADIO

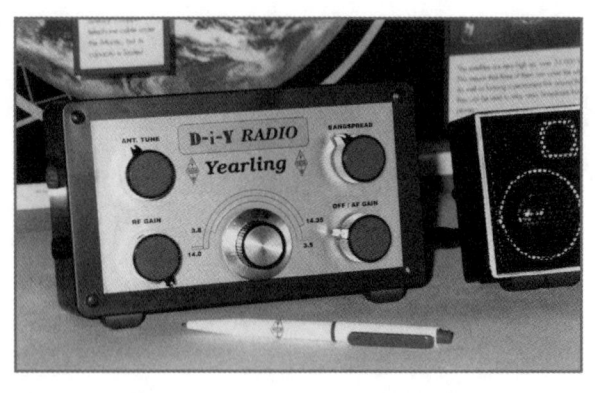

**Photo 4.3: The Yearling Receiver**

er to implement, and as a result many more amateur HF (and other) receivers and transceivers are using DSP back-ends. Normally, it is only used to replace the analogue IF and demodulator stages. However, there are also some external DSP filters that can be added to various sets. These audio only DSP units are not as versatile as those that take an IF signal. The audio processors are only able to provide audio filtering, whereas "RF DSP" units are able to take the IF signal, provide filtering at the intermediate frequency as well as demodulation and any audio filtering that may be required.

## Receiver project

The Yearling easy-to-build receiver design by Paul Lavell, G3YMP, was first published in D-i-Y Radio and later in RadCom. The original design was for the 20m band but it has been extended to cover 80m as well. The receiver is powered from either a 9V PP3 battery or mains adapter and can be built either on the printed circuit board (PCB) supplied with the kit or on a prototype board. The building instructions here are based on the PCB and kit. Either headphones or a loudspeaker can be used.

## Circuit design

Direct conversion was the first option considered for the design of the Yearling. This type of receiver has the merit of simplicity but, unless carefully designed and constructed, can suffer from problems such as strong breakthrough from broadcast stations on nearby frequencies. Also, a stable VFO at 14MHz, while certainly not impossible, is not something to be undertaken lightly by a beginner. The superhet was then considered. This overcame some of the problems associated with direct conversion but created others such as the need for a rather expensive IF filter. So the result was a happy compromise, which in effect is a direct-conversion receiver preceded by a frequency converter. This means that the VFO runs at about 5MHz instead of 14MHz, so stability is much better.

The circuit is given in Fig 4.18. Incoming signals at 14MHz or 3.5MHz are selected by the tuned circuit L1/C1/D1 or filter FL1. Note that the ANT TUNE control is not needed on 80m as FL1 provides the necessary filtering. Tuning on 14MHz is carried out by RV2 which adjusts the voltage on varicap D1. This is one half of diode type KV1236 - note carefully the polarity of this component.

A Philips SA602 (IC1) converts the signal to the range 5.0 to 5.5MHz (approx) by means of its internal oscillator. This has a crystal (X1) working at about 8.9MHz. In fact, any crystal between 8.8 and 9.0MHz will be satisfactory but a frequency at about 8.95MHz will give greatest accuracy. It should be noted that D1 is in fact forward biased over part of its voltage range. However, the circuit as it stands performs quite adequately.

The signal output from the mixer in IC1 then passes to the IF filter formed

# CHAPTER 4: RECEIVERS

**Fig 4.20: Circuit diagram of the Yearling Receiver**

by C5 and L2. The tuned circuit is damped by the rather low output impedance of IC1 and this gives a nice compromise between selectivity and insertion loss. The balanced output of the tuned circuit is applied to IC2, another SA602 mixer/oscillator which acts as a product detector.

The main VFO uses the oscillator section of IC2, which covers a range of approximately 5.0 - 5.5MHz. Assuming the use of a 9MHz crystal for X1, the 20m band will track within the range 5.00 - 5.35MHz and the 80m band from 5.2 - 5.5MHz. Note that the LF ends of the respective bands will be at opposite ends of the dial, since 20m makes use of the sum of the receiver's two oscillator frequencies, and 80m uses the difference between them.

Main tuning is carried out by RV4, and RV5 provides the bandspread control. Tuning is by means of the voltage on varicap D2 which, in association with C9, C10 and L3, determines the frequency of oscillation. The varicap is a dual type - cut in two with a sharp knife. Voltage regulator IC3 in the supply lines to the early stages makes stability surprising good. The audio output from IC2 is amplified by IC4a and filtered by low-pass filter IC4b, before being further amplified to speaker level by IC5.

## Construction

Many Yearlings have been built from kits without problems, while a number of constructors have successfully used a prototype board instead of the PCB. In fact, the initial design for the receiver was built using a matrix or prototype board. No special precautions are needed for the construction but, as with any radio, neat wiring makes the tracing of faults much easier.

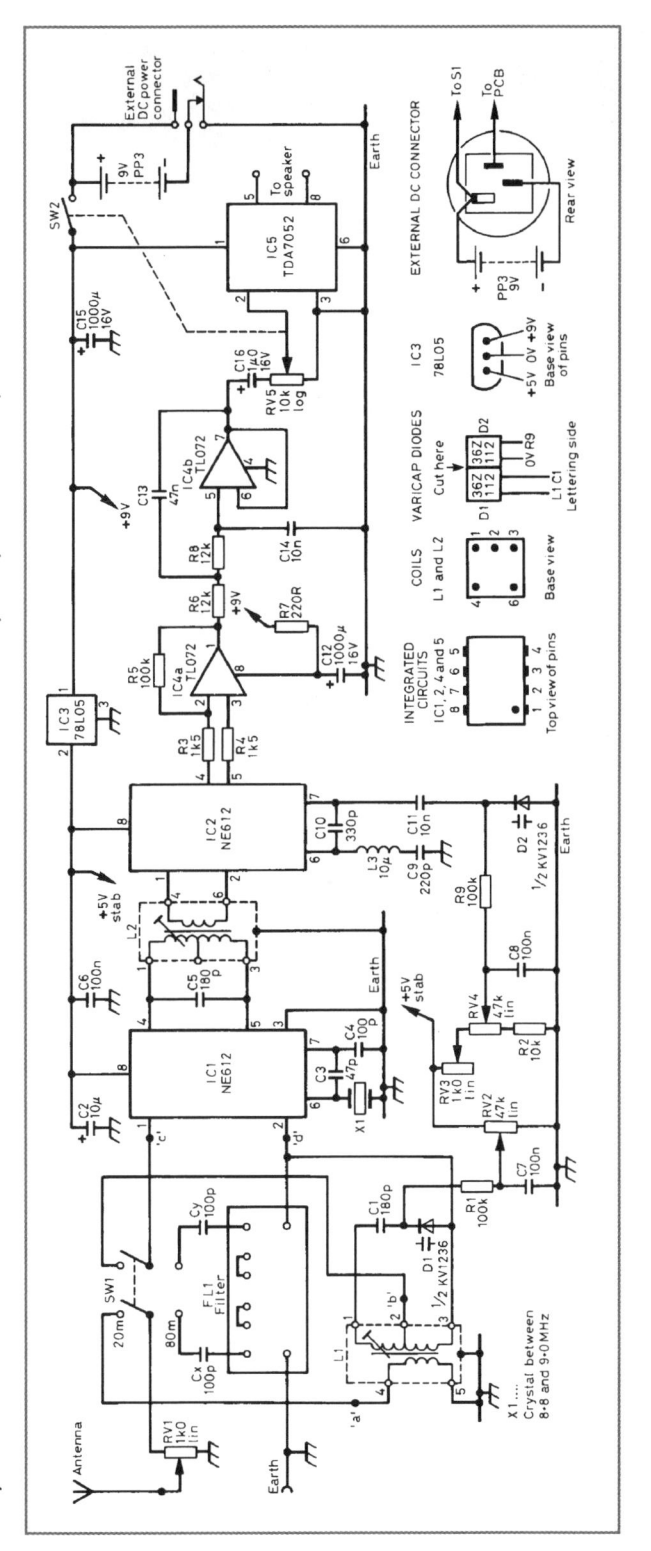

# HF AMATEUR RADIO

Fig 4.21 shows the connections to the gain and tuning controls. Screened cable should be used for the leads to the volume control but stranded bell wire should be satisfactory elsewhere. Incidentally, there was no problem with using IC sockets for all the 8-pin devices. The coils are colour coded, with L1 having a pink core and L2 a yellow one.

It is rather easy to wire the varicaps incorrectly but, if using the PCB, the lettering on D1 should be next to coil L1 and the lettering on D2 should be facing resistor R7. Fig 4.22 shows the band-change switch and the 80m filter which is glued to the side of the case. Holes for the five controls are 10.5mm diameter, and the speaker and power connectors have 6.3mm and 11mm holes respectively. The antenna and earth sockets need 8mm holes.

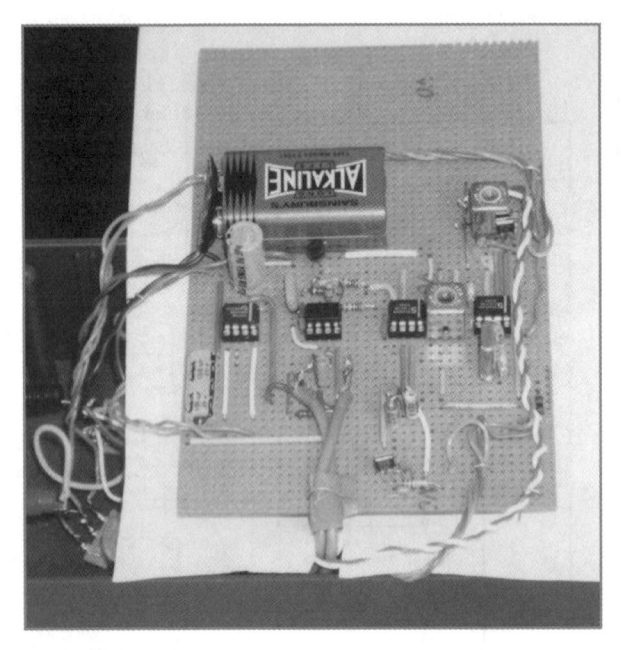

Photo 4.4: The Yearling may also be built on prototype board

## Setting it up

Connect a 9V battery and a reasonable antenna, and, on switching on, some stations - or at least some whistles - should be heard. It is suggested that a start is made with the 20m band, and the adjustments made before fitting the controls and sockets to the case.

A signal generator is useful, of course, but not essential to get the receiver working. The following steps should be followed, whereupon the Yearling should burst into life.

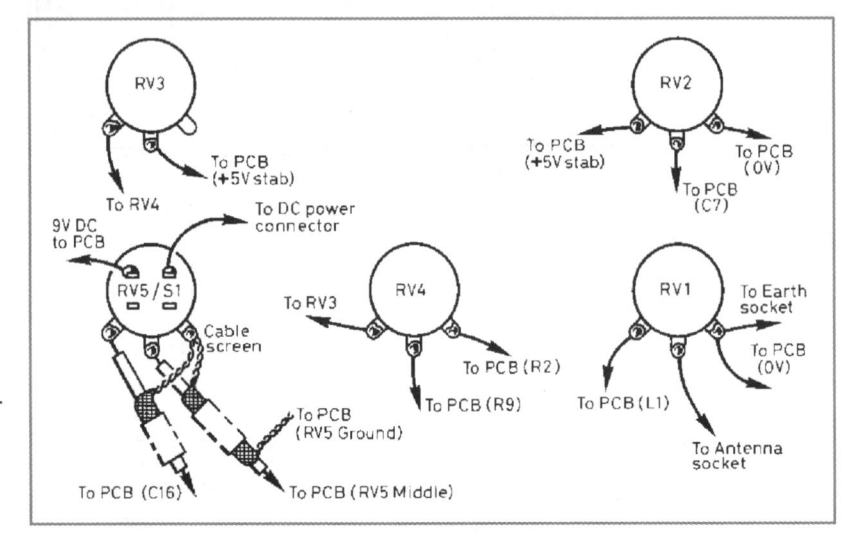

Fig 4.21: Rear view of variable resistors. Check the connections carefully to make sure the wires fit to the correct holes on the board

# CHAPTER 4: RECEIVERS

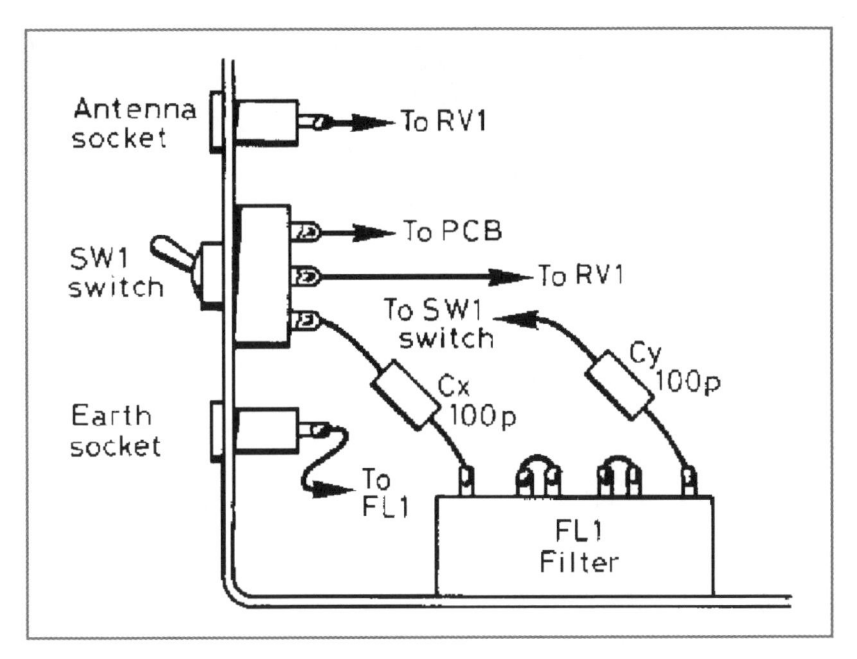

**Fig 4.22:** Internal view of the Yearling case. The 80m filter is attached to the base using glue

1. Set the core of L2 to mid-position.
2. Set RV1, RV2 and RV4 to mid-position and rotate the core of L1 until you hear a peak of noise. Now adjust L2 for maximum noise.
3. Tune carefully with the main tuning control RV4 until you hear amateur signals. Adjustment of the bandspread may be needed to clarify the speech.
4. Switch off the receiver and fit the controls and sockets to the case.
5. Finally, adjust the tuning knob so that the pointer roughly agrees with the dial. Due to the spread of varicap capacitance values, you may find the tuning a little cramped. This is easily fixed by adding a resistor (try 22kohms to start with) in series with RV3 or adjusting the value of R2.
6. Check the 80m band - this should work without further adjustments to the coils. Fig 4.20 shows the additional connections for 80m as the Yearling was originally designed for 20m only.
7. Finally, fix the PCB inside the case (double-sided sticky tape works well).

**Fig 4.23:** The underside of the PCB. Wires are connected from the switch and filter as shown

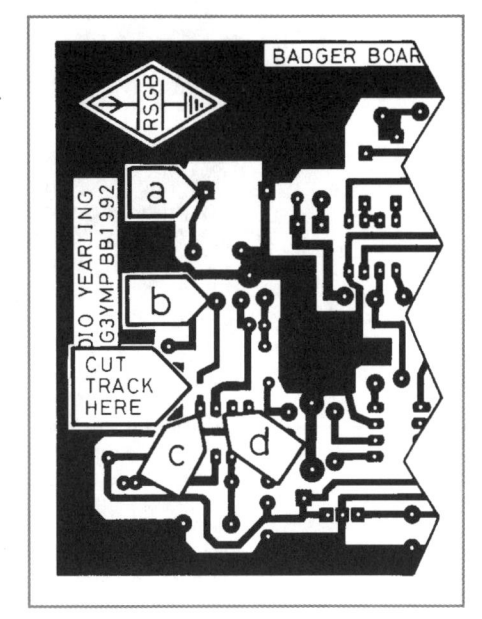

# HF AMATEUR RADIO

### Table 4.1: Parts List

**Capacitors**
All rated at 16 V or more

| | |
|---|---|
| C1, 5 | 180p polystyrene, 5% or better |
| C2 | 10µ electrolytic |
| C3 | 47p polystyrene, 5% or better |
| C4, Cx, Cy | 100p polystyrene, 5% or better |
| C6 – C8 | 100n ceramic |
| C9 | 220p polystyrene 2% or better tolerance |
| C10 | 330p polystyrene 2% or better tolerance |
| C11, 14 | 10n ceramic |
| C12, C15 | 1000µ electrolytic |
| C13 | 47n 5% polyester |
| C16 | 1µ electrolytic |

**Resistors**
All 0.25W 5%

| | |
|---|---|
| R1, 5, 9 | 100k |
| R2 | 10k |
| R3, 4 | 1k5 |
| R6, 8 | 12k |
| R7 | 220R |
| RV1, 3 | 1k linear |
| RV2, 44 | 7k linear |
| RV5 | 10k log with switch (SW2) |

**Inductors**

| | |
|---|---|
| L1 | Toko KANK3335R |
| L2 | Toko KANK3334R |
| L3 | 10µ, 5% tolerance (eg Toko 283AS-I00) |

**Semiconductors**

| | |
|---|---|
| IC1, 2 | Philips SA602 |
| IC3 | 78L05 |
| IC4 | TL072 dual op-amp |
| IC5 | Philips TDA7052 audio amp |

**Additional items**

Varicap diode, Toko KV1236 (cut into two sections)
Crystal between 8.8 and 9.0MHz. An 8.86MHz type is available from JAB, Maplin, etc
Wavechange switch, DPDT changeover type
8-pin sockets for IC1, 2, 4 and 5
4mm antenna (red) and earth (black) sockets
3.5mm chassis-mounting speaker socket
DC power socket for external power (if required)
4 knobs, approx 25mm diameter with pointer
Tuning knob with pointer, e.g. 37mm PK3 type
Printed circuit board or prototype board
Plastic case, approx 17 x 11 x 6cm, eg Tandy 270-224.
Speaker between 8 and 32 ohms impedance (or headphone)

*A kit of parts is available from JAB Electronic Components, The Industrial Estate, 1180 Aldridge Road, Great Barr, Birmingham B44 8PE, email jabdog@blueyonder.co.uk.*

CHAPTER 5

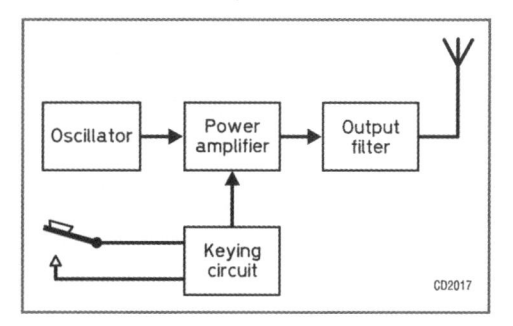

# Transmitters

*In this chapter:*

- QRP transmitters
- High-power transmitters
- Linear amplifiers
- VOX and break-in
- Transmitter specifications
- Speech processing
- Transceiver project

THERE are many transmitters from which to choose these days. Often they are part of a transceiver because this gives the maximum flexibility and ease of use. But even in such cases, it is important to pay serious consideration to the way the transmitter operates and and its specifications. As with receivers, you need to understand these specifications buying a unit to make sure it meets your requirements, and an appreciation of how a transmitter works makes it possible to use the equipment more effectively and reliably.

## QRP-style transmitters

Today a large number of people like to venture onto the bands with low-power equipment they may have built themselves. There is a great feeling of achievement in making contacts with homebrew equipment. The transmitters used are often relatively simple, and many only cater for Morse. These simple transmitters can easily be built by most amateurs. However there are also many QRP SSB transceiver projects and products available that provide high level performance, but using low power, and which are highly complicated to build.

A simple Morse transmitter can be made using as few as three transistors and a handful of other components. Typically, one transistor would be used as a crystal oscillator and the second as the power amplifier as shown in the block diagram of Fig 5.1. The third transistor is often used to switch the PA and prevent the key from having to pass the full PA current which might cause pitting or burning of the key contacts. While a crystal oscillator restricts the range of transmission frequencies you can use, it simplifies the circuit considerably. A variable-frequency oscillator (VFO) requires the use of several more transistors and this complicates the construction. However, a VFO does enable the fre-

Fig 5.1: Block diagram of a simple QRP CW transmitter

# HF AMATEUR RADIO

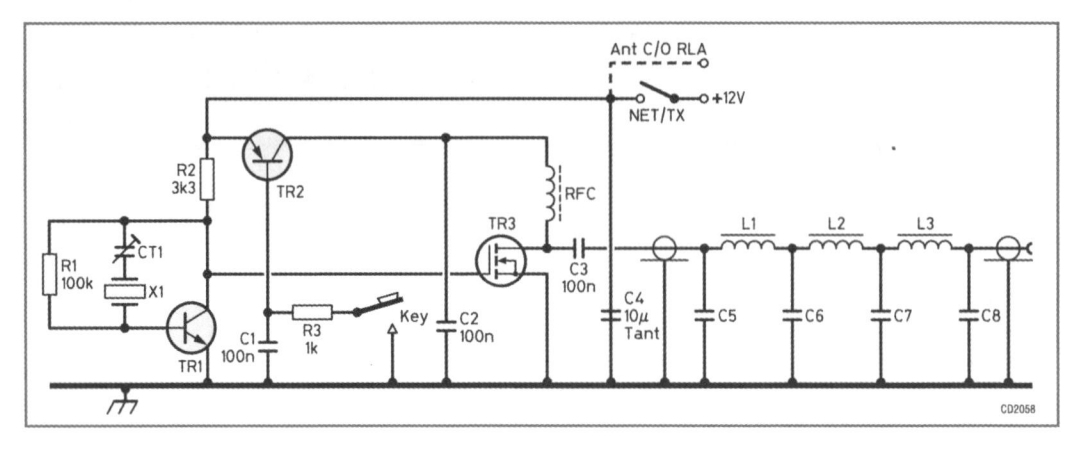

**Fig 5.2: Circuit of a typical QRP transmitter**

quency of operation to be chosen. This can be a great advantage when running QRP as it means that other stations can be called rather than having to rely calling CQ, or hoping that other people just happen to be on the same frequency as your crystal.

On the output of the transmitter, there is a filter and matching network. This provides two functions. The first is to provide a good match between the output transistor and the antenna, thereby ensuring that the maximum amount of power is transferred into the latter. The second function is to filter any out-of-band spurious products. Any amplifier, and particularly one running in Class C, will produce harmonics, and these need to be re-moved to ensure they do not cause interference. Typically any spurious signals should be about 50dB below the carrier level.

Transmitter designs are regularly published in amateur radio magazines. Alternatively, there is a growing number of companies that sell kits, which can often be completed in an evening or two. In many cases, metalwork can also be bought with the kit, and this provides a very neat case for the unit, giving it a professional look.

Usually, QRP transmitters run at powers of 10W and below. They can be great fun to build and use. However, the very fact that they run low levels of power means that contacts will not be as plentiful as they are for those who have higher-power transmitters. Accordingly, time should be spent in making the antenna as efficient as possible.

**Photo 5.1: A kit QRP transmitter. These kits are often very easy to build and provide a very cost-effective way of putting a signal on the air**

# CHAPTER 5: TRANSMITTERS

## Single sideband transmitters

While there is a large group of QRP operators and constructors, most people tend to use higher-power, multimode transmitters that are part of a transceiver. These normally transmit either Morse or single sideband, and some even have frequency-modulation capabilities.

The way in which the signals are generated is very much the inverse of how a superhet receiver works. This means that most of the circuits used in the receiver can also be used in the transmitter section, preventing duplication of many areas of circuitry. However, for simplicity's sake, we will now take a look at circuit for a standalone transmitter. Exactly the same principles hold for the transmitting sections of a transceiver.

The block diagram of a basic transmitter is shown in Fig 5.3. First, the carrier is generated using a crystal oscillator. Often a frequency in the region of 8 to 10.7MHz is chosen. This is high enough to simplify the problem of removing unwanted mix products later when the sideband signal is converted to its final frequency. This is particularly important when the transmitter forms part of a transceiver and unwanted mix products may appear in the receiver as 'birdies'. Another advantage of using an IF in this frequency range is that it is not so high that suitable filters are prohibitively expensive.

The signal from the carrier oscillator is applied to a balanced modulator, together with the amplifier and processed audio. The action of this modulator (mixer) produces a signal containing two sidebands (the sum and difference frequencies). A double-balanced mixer is normally used and this means that the input signals are suppressed at the output, giving a double sideband, suppressed carrier signal. This signal is then applied to a sideband filter to remove the unwanted sideband, leaving a single sideband, suppressed carrier signal at the intermediate frequency.

With the basic signal generated, the next stage is first to move it to the required frequency and then amplify it to the required level. Conversion to the right frequency may be undertaken in two or even three stages, as it is in many receivers. One of the mix processes will involve the use of a variable-frequency oscillator. Today, this is most likely to be a synthesised oscillator, the frequency and operation of which is under microprocessor control. This gives the opportunity to have facilities like dual VFOs and the like, even though only one synthesiser is used. After each mix stage, careful filtering is required to ensure that the level of spurious signals is kept acceptably low. Once the signal is on the right frequency, it is amplified to bring it to the required power level. For most transceivers today, this is around 100W output.

Fig 5.3: The block diagram of an elementary SSB transmitter

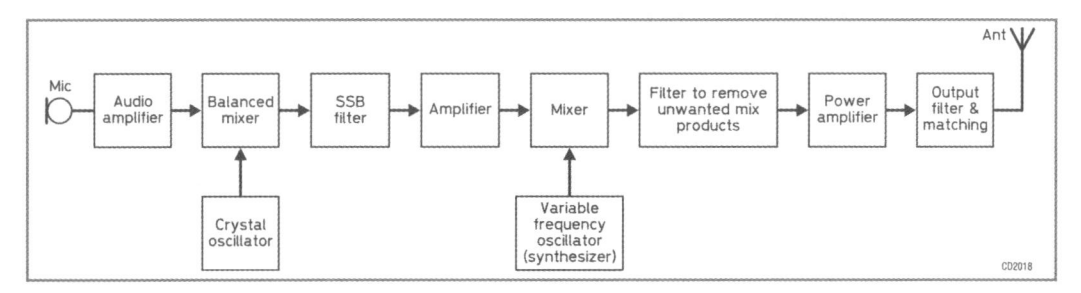

# HF AMATEUR RADIO

**Photo 5.2: A typical HF transceiver**

QRP transmitters require filtering and a matching network at the output of the power amplifier, and the same is true for higher-power Morse and single sideband transmitters. In fact, the level of spurious signals is even more important because power levels are generally much higher and the possibility of causing interference is that much greater.

## VOX and break-in

A system known as press-to-talk is normally used to change a transmitter from receive to transmit. This is normally accomplished with a pressel (press switch) on the microphone. (When using Morse, a transmit/receive switch can be employed.) However, under many circumstances, it is convenient to be able to speak and let the electronics detect the presence of sounds, automatically switching from receive to transmit. This system, known as VOX, is included on many sets.

While the system may appear to be very convenient, it is not quite as foolproof as it may appear. First, any noise of a sufficient level may trigger the system and this can result in the transmitter changing over at the wrong time. There is normally a control to set the level when the changeover takes place but even this does not give complete immunity from unwanted changeovers. Second, even sounds from the receiver loudspeaker can find their way into the microphone and cause the transmitter to transmit. This can be overcome with the 'Anti-vox' control. This effectively nulls out the receiver sounds so that they do not affect the vox operation. Finally, it takes a finite time for the receive/transmit changeover to take place. If relays are used in the changeover, the time taken may result in the first syllable being clipped. This can often be detected at the receiving end, and can sound a little odd. As a result, most DX operators prefer to use press-to-talk and have the operation completely under their control, as unwanted or spurious changes can result in them missing receiving important information.

For Morse operation the situation is a little different. Virtually all key presses are likely to be intentional and as a result break-in, as it is called, is far more useful. However, it is very important that changeover is fast, otherwise the first dot or dash will be shortened. Under some circumstances, when high speeds are used and the changeover is slow, it can result in the first dot being missed. Some transceivers use all-electronic switching in the changeover and can accomplish the switch from transmit to receive and back again very quickly. In these instances the response of the receiver AGC to the changes is the main factor. When this is well designed, full break-in is very useful, offering the ability to transmit by simply pressing the key. It also enables listening between the dots and dashes, giving a much greater 'visibility' of what is happening while you are transmitting. This can be an enormous advantage when many people may be calling a DX or contest station at the same time.

## Keying

Most new transceivers contain keyer electronics and simply require a paddle to be connected to provide a fully iambic-mode keyer (i.e. a fully automatic keyer that generates a sequence of interspersed dots and dashes when both paddles are squeezed together).

## Speech processing

It has been said that single sideband without the use of speech processing is an outmoded form of communication. The reason for this is that the output from a single sideband transmitter is limited by the peak audio level that is being transmitted. If the average power is very low, then the signal will be perceived as being relatively weak. If the average power of the signal can be raised then the signal will appear to be much stronger and will be able to be heard more easily through interference and above the noise. Unfortunately, speech has a very high content of short-lived peaks and a low average power level. To make the best use of the transmitter power, the average level of the audio has to be raised using a speech processor.

There are a number of reasons why speech has a low average power. First, the intensity of the speech may vary – the speaker may place more emphasis on one word than another or may move away from the microphone. In this way, the overall level of the speech varies. This can be very significant and even when the speaker is trying to maintain a constant level it can vary by 10dB. Another reason for the low average level is that speech contains short-lived peaks or transients.

Sounds like 'p' and 'b' are known as plosive sounds and start with a burst of energy after which the level falls away. Other sounds too, have large peaks, whereas the vowel sounds tend to be more constant.

In addition, not all frequencies contribute as much towards the intelligibility of a signal. The best use can be made of the available power by ensuring that only the frequencies that contribute most to the intelligibility of the signal are included.

**Fig 5.4: A typical speech waveform, showing that human speech has a very low average power. As the transmitter is limited by the peak level it can carry, the average level of the modulation must be increased to use the available power capability more efficiently**

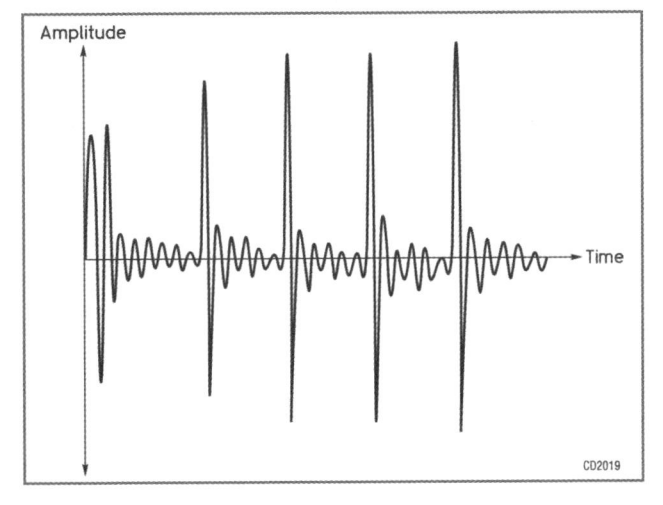

# HF AMATEUR RADIO

Photo 5.3: A typical example of an external speech processor

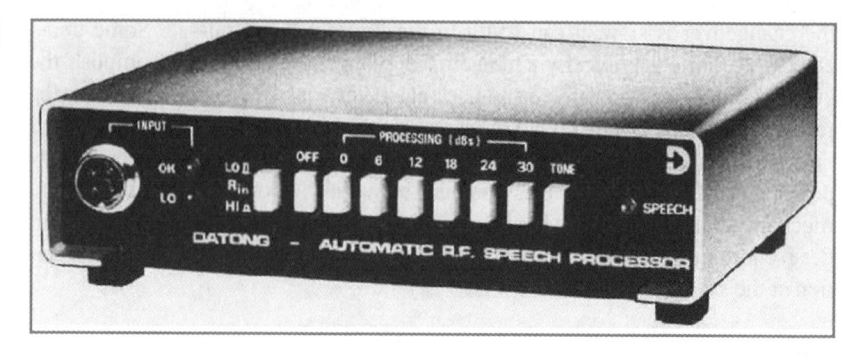

Fig 5.5: Block diagram of a compressor using a time constant. This type of device is often called VOGAD (voice-operated, gain-adjusting device) and it is used to maintain the general level of the signal

Processing is achieved in a number of ways: compression, clipping and frequency tailoring all play their part. Some of the more sophisticated processors use all three methods and can give up to 8dB gain – more than most linear amplifiers can supply and at much less cost.

Speech compression is a form of processing where the gain of an amplifier is steadily reduced as the audio level increases. Generally, the gain reductions only occur once a certain level is reached. The gain may be reduced instantaneously so that it varies over the audio waveform itself.

Another form of compression has a time constant introduced so that it acts like an audio AGC and alters the gain according to the peak level of the signal (Fig 5.5). In this way, the general level of the signal is maintained at a given level. This type of circuit is often called VOGAD (voice-operated, gain-adjusting device). The time constants in the circuit are important. The attack time should be fast to ensure that any sudden increases in level do not pass beyond the compressor and cause any overloading – typically speeds of around 10ms are used for this. The decay time should be much slower so that the gain level follows the level of the speech – usually decay times of around 300ms are employed.

Clipping may also be used (Fig 5.6). This technique involves removing the peaks of the waveforms. In this way, the peak-to-average ratio can be considerably improved. While it may appear that removing part of the wave-form might remove vital information from the waveform and reduce the level of intelligibility, this is not the case if the clipping is performed well. The reason for this is that the sounds are recognised primarily by their frequency content.

The level of clipping is important when clipping a signal. A figure is often quoted in decibels and this refers to the ratio between the peak levels of the signal before and after clipping.

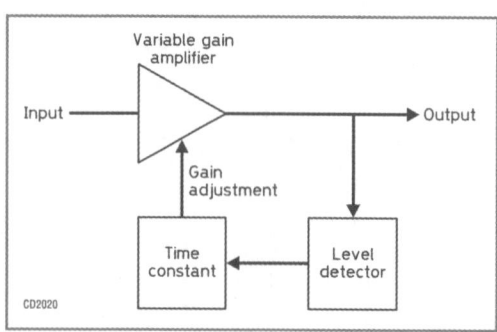

Unfortunately, the clipping process is not linear, and distortion products are introduced into the signal, most of these being in the form of harmonic distortion (Fig 5.7). Typically, signals for communications purposes are limited to 3kHz, and these distortion products will cause the frequency content from the clipper to extend beyond this limit. This means that a clipper of this type would need to have a filter after it to remove these products. Unfortunately, many frequencies in the wanted signal give rise to

60

harmonics that fall within the wanted audio bandwidth, and these have the effect of reducing the level of intelligibility. As a result, the level of clipping suitable for 'baseband' (audio) clipping is limited to about 10 to 15dB – beyond this, the level of distortion starts to reduce the intelligibility of the signal. This limits the actual gain that can be achieved by an audio processor to around 5dB.

To overcome the problem, a radio frequency signal can be used. When this is clipped the harmonics occur at multiples of this signal and can be easily filtered out (Fig 5.8). To achieve this, a single sideband signal must be generated. A signal with good sideband and carrier suppression is required to reduce intermodulation products that would impair the overall performance. The signal is clipped at the radio frequency, filtered to remove the out-of-band products including harmonics and then demodulated to regenerate the clipped audio. In some transmitters, the clipper is included in the sideband generator to provide an integrated processor within the transmitter. In this case, it is not necessary to include the demodulator stage of the processor.

While RF clippers are obviously more complicated than their AF counterparts they can produce significantly more gain. As already mentioned, the maximum gain that can be usefully obtained with a AF processor is around 4 to 5dB, whereas an RF processor can give up to about 8dB. Every decibel of gain is useful in helping to make the station more effective and make contacts through pile-ups or under low-signal conditions.

Useful improvements can also be made by ensuring that the frequency content of a signal is tailored. Even landline telephone systems have their frequency response limited to cover frequencies between 300 and 3000Hz. Communications systems normally tailor this even more, limiting the top frequency to 2.7kHz or even slightly less. While this does reduce the natural sound of the signal, there is only a marginal drop in intelligibility.

Reducing the low-frequency content of the signal and pre-emphasising the higher frequencies is found to reduce an effect during clipping where the stronger low-frequency components are emphasised, making the signal sound rather unnatural. Typically, a response that falls below 1kHz may be useful.

Many processors today combine all three methods (Fig 5.9). On entry to the processor, the signal may have its frequency bandwidth limited and the low-frequency content reduced. The signal may then enter a VOGAD stage to ensure a constant audio level is presented to the clipper. After clipping, preferably at RF, the signal is filtered and in the case of an RF filter demodulated before being presented to the transmitter audio input.

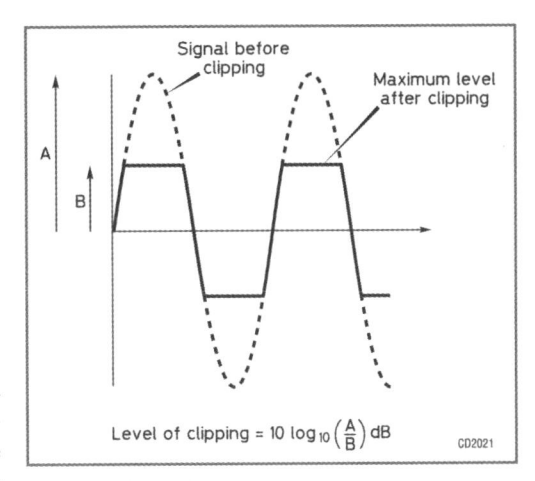

Fig 5.6: The level of clipping is the ratio between the peak levels before and after clipping expressed in decibels

Level of clipping = $10 \log_{10}\left(\frac{A}{B}\right)$ dB

Fig 5.7: Harmonic distortion products created during audio clipping. Those above the maximum required frequency can be removed by filtering, but those in the required audio band-width cannot be re-moved and reduce the intelligibility of the signal

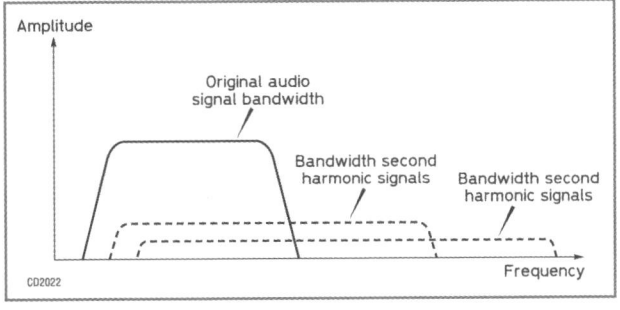

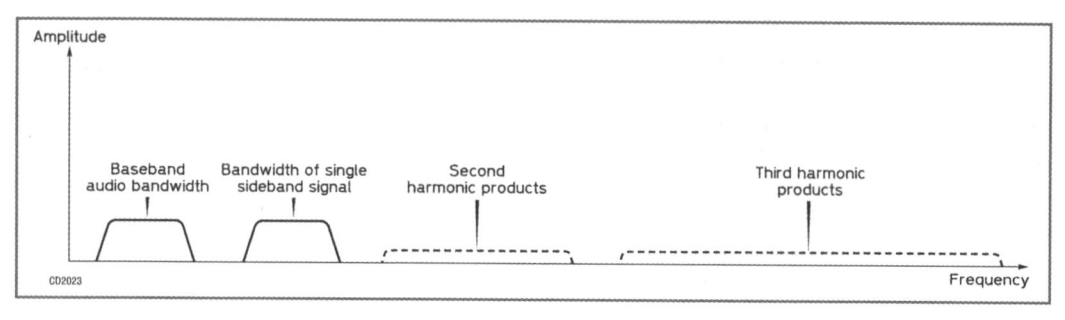

**Fig 5.8: By clipping a radio frequency signal, the harmonics caused by clipping fall at multiples of this signal and can be filtered out very easily, thereby reducing the level of distortion and allowing greater levels of clipping to be achieved**

Processors are particularly useful because they enable the best use to be made of the available RF. In some cases, they may be used instead of a linear, providing a very cost-effective solution.

## Linear amplifiers

Usually HF transmitters are capable of providing around 100W output power. In most countries, the maximum legal limit is somewhat higher and to reach it a further amplifier is needed. Amplifiers used for single sideband must be linear – if they are not, the waveform becomes distorted and the signal will splatter over the band, causing interference to other users.

As amplifiers are often manufactured so they can be used anywhere in the world, it is possible that they can produce power in excess of the legal limit for the country where they are being used. Accordingly, care should be taken to ensure this limit is not exceeded.

While linear amplifiers can be expensive, they can provide a useful amount of gain. In the UK, the maximum output power is 400W (26dBW), and in many other countries maximum power limits of up to 1 kW (30dBW) are permitted. When compared to a transmitter with an output of 100W, the station using the linear at 400W will be 6dB higher in level. This is a significant amount and could make the difference between making a contact or not. While a linear is not an essential piece of equipment, it can certainly make a difference, especially when the going gets tough.

**Fig 5.9: A speech processor combining several forms of processing to give the optimum performance**

## Transceivers

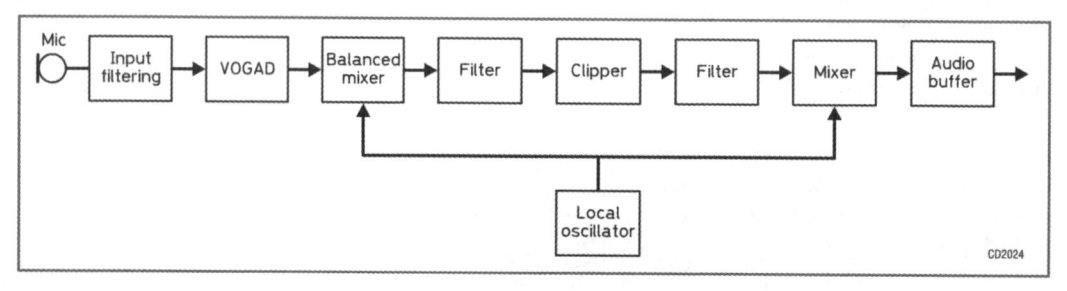

Most people do not buy individual receivers or transmitters, but transceivers containing all the circuitry for transmitting and receiving. Obviously, a single unit is far more convenient than having separate units. Also, the cost of a transceiver is less than the sum of its individual counterparts because a large portion of the circuitry can be used for both transmit and receive functions. Stages like the mixers, local oscillator, IF amplifier and so forth can all be used for both receive and transmit.

In order to make a transceiver, the routing of the signal has to be changed between transmit and receive. This can be done using either relays or semiconductor switches such as PIN diodes.

## Transmitter power

There are a number of methods of making power measurements. Some years ago, it was common to measure the DC power input to the final amplifier. This method was adopted because it was easy to measure the voltage supplied to the final amplifier and the current drawn. It was also easy because the current was steady for Morse and the AM transmissions that were being used at the time.

These days, power is normally measured at the output of the transmitter. While many measurements are made directly in watts, it is more common to define power in terms of 'dBW', and indeed the UK licence quotes power levels in these terms. Although dBW may not be quite as easy to use, it is quite straightforward to understand. It is simply the power expressed in decibels relative to one watt. To give an example, a power of 10W is 10dB increase on 1W, and so it can be described as a level of 10dBW. Similarly a power of 400W is 26dB above 1W and this means it can be expressed as 26dBW.

While it is relatively easy to measure the power of a steady waveform such as that of a Morse signal with the key held down, the same is not true of a single sideband signal. The output power varies in line with the instantaneous level of the audio. The power of a single sideband signal is therefore measured at the peak of the radio frequency signal envelope. Accordingly, this power is called the peak envelope power (PEP).

## Spurious signals

With many HF transmitting stations radiating significant levels of power, it is necessary to ensure that they do not generate any undue levels of spurious signals that

Photo 5.4: A typical example of a linear amplifier

might cause interference to other users. There are a number of causes of spurious signals – one is when harmonics are generated, another arises from intermodulation distortion, and a third is caused by unwanted mix products being insufficiently filtered and thus being present at the output.

## Intermodulation

One of the major causes of this type of interference is poor linearity of amplifier stages in the transmitter, and in particular the final amplifier. This gives rise to intermodulation products in the same way as in a receiver front-end amplifier. The products arise when two or more signals are passed through the amplifier. While this distortion will not occur in the case of a Morse signal where only one carrier is present, a single sideband signal consists of a whole variety of different frequencies within the transmitted bandwidth.

When specifications are given, they take the case where two tones or closely spaced signals are passed through the amplifier. To illustrate how this is done, imagine a single sideband signal derived from just two audio tones, one at 1kHz and the other at 2kHz. The resulting sideband signal will be two radio frequency signals spaced by 1kHz. When passed through an amplifier these will give rise to third-order and higher-order intermodulation signals in the same way that was described for a receiver amplifier in Chapter 4. In the case of the receiver, it was seen how the third-order mix products gave signals at $2f1 + f2$, $3f1 + 2f2$, etc and $2f2 + f1$, $3f2 + 2f1$ etc as shown in Fig 5.10.

In the case of a real single sideband signal, there will be a whole variety of different audio frequencies. All these intermodulate with one another to generate noise or splatter which spreads out from the main signal. Normally, the worst intermodulation products will be those nearest the wanted signal, and their levels reduce as the offset increases.

The specifications for intermodulation products are usually given in terms of the difference (in decibels) between the wanted or main signal and the various intermodulation products. However, a transmitter specification will often say that all intermodulation products are below a given level, and in this case the worst ones are bound to be the third-order products. Sometimes (especially in a review) the levels of specific products will be stated.

Fig 5.10: Intermodulation products generated in a linear amplifier

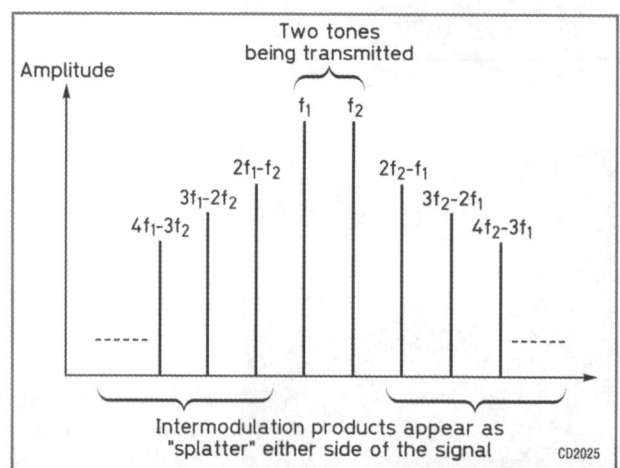

Typically, values are around 25 to 30dB for the third-order products and five or six decibels lower in the case of the fifth-order products (the more negative the number, the better the performance). A typical modern transmitter should have all its products better than 25dB relative to the main signal.

## Harmonics and other spurious signals

Harmonics of the required frequency are also generated in a transmitter. Other signals may also appear at the output. These may arise from the mixing process in the transmitter. Whatever the cause, they need to be kept to acceptable levels to ensure that no undue interference is caused to other users.

The level of a spurious signal is not normally given in absolute terms, i.e. a given number of watts or milliwatts. The more usual way is to relate it to the level of the wanted signal. In other words, a spurious signal will be said to be a certain number of decibels below the carrier, sometimes measured in 'dBc' units. Again, the more negative the number is, the smaller the spurious signal level. For example, a spurious signal level of 40dBc is better or lower in level than one of 31dBc.

The ideal level for spurious signals would be zero but this is obviously impossible. Normally, transmitters are specified as having all spurious signals at least 40dB below the carrier, i.e. 40dBc. Often the specification is left rather vague because the levels of harmonics and other spurious signals will vary according to the band or frequency in use. Typically harmonics may be at levels of 50dBc or even as much as 60dBc.

In cases where interference is being experienced on frequencies above 30MHz, it may be necessary to add a low-pass filter to an HF transmitter. These filters are available from many amateur radio stockists and give very high levels of out-of-band attenuation. If fitted correctly, they will reduce any above-band harmonics to well within acceptable limits.

## VSWR tolerance

When transmitters operate with high levels of standing waves, high voltages may be present at the power amplifier. These could cause the output devices to operate outside their ratings and may result in their destruction. To prevent this happening, most transmitter power amplifiers incorporate protection circuitry. When a high level of VSWR is detected, this circuitry limits the output from the transmitter to a safe level. Generally, a VSWR level of 2:1 is acceptable but 3:1 may cause a power reduction to be noticed.

## Incremental tuning and separate VFOs

Most transceivers today have very flexible tuning facilities. The exact way these operate depends on the particular transceiver in question and may include incremental receiver tuning (IRT) and multiple VFOs.

Incremental receiver tuning is normally used to give a small amount of receiver shift while retaining the transmitter on the same frequency. This is very useful when needing to 'clarify' the incoming signal by changing the receiver tuning very slightly. If the main tuning control were used, then it would change the transmitter frequency as well. The receiver incremental tuning control is generally an additional knob that normally gives a shift of a few kilohertz at the most.

The majority of modern transceivers have a twin VFO facility. This enables the tuning to be set to one channel on one VFO, and another channel on the other VFO. Switching between the two VFOs is normally very easy, and almost invariably it is possible to receive on one VFO and transmit on the other. In this way, split-frequency operation can be undertaken. This is particularly useful because many DX stations transmit on one frequency and receive on another a few kilohertz off-channel to ease operation and manage a pile-up when many stations are calling. Typically, the transmit and receive frequencies may be between 1 and

5kHz apart when using Morse and between 5 and 20kHz apart when using sideband. This degree of shift is often beyond the scope of the incremental receiver tune facility and in such case a twin VFOs is far easier to use and more flexible.

## Automatic level control

As the intermodulation performance depends largely on the linearity of the final amplifier, it is imperative that the latter is not over-run. When this happens, it will start to limit and there will be a marked increase in the distortion and hence the amount of 'splatter' being picked up by nearby receivers. To prevent this, transmitters use automatic level control (ALC) to detect the level of the signal at the output and reduce the gain of previous stages to prevent overload. In fact, it operates in the same way as AGC in a receiver. The main point to note is that no transmitter amplifier should be run close to its limits, otherwise distortion levels rise and cause interference to other users.

## Practical design of a QRP Transceiver with full break-in

This design by Peter Asquith, G4ENA, originally appeared in Radio Communication May 1992. The QSK QRP Transceiver block diagram is shown in Fig 5.11. The circuit employs several digital components which, together with simple analogue circuits, provide a small, high-performance and low-cost rig. Many features have been incorporated in the design to make construction and operation simple.

One novel feature of this transceiver is the switching PA stage. The output transistor is a tiny IRFD110 power MOSFET. This device has a very low 'ON' state resistance which means that very little power is dissipated in the package, hence no additional heatsinking is required. However, it does mean that good harmonic filtering must be used.

The HC-type logic devices used in the rig are suitable for operation on the 160m and 80m bands. The transmitter efficiency on 40m is poor and could cause overheating problems. Future advances in component design should raise the top operating frequency limit.

The circuit diagram of the transceiver is shown in Fig 5.12. The VFO is based around TR2 which is used in a Colpitts configuration to provide the oscillator for both receive and transmit.

The varicap diode D1 is switched via RV2/R1 by the key to offset the receive signal by up to 2kHz, such that when transmitting, the output will appear in the passband of modern transceivers operating in the USB mode. C1 controls the RIT range and C29/30 the band coverage. IC1a, IC1b and IC2a buffer the VFO. It is important that the mark/space ratio of the square wave at IC1b is about 50:50. Small variations will affect output power.

For the transmitter, when the key is operated, RL1 will switch and, after a short delay provided by R19/C15, IC2c and IC2d, will gate the buffered VFO to the output FET, TR3. TR3 operates in switch mode and is therefore very efficient. The seven-pole low-pass filter after the changeover relay removes unwanted harmonics, which are better than -40dB relative to the output.

The receiver circuit is straightforward. The VFO signal is taken from IC2a to control two changeover analogue switches in IC5, so forming a commutating

**Photo 5.5: The G4ENA transceiver**

mixer and providing direct conversion to audio of the incoming stations. IC4a is a low-noise, high-gain, differential amplifier whose output feeds the four-pole CW filter, IC4a and IC3a, before driving the volume attenuator, RV1. IC3b amplifies and TR1 buffers the audio to drive headphones or a small speaker. When the key is down TR4 mutes the receiver, and audio oscillator IC1a and b injects a sidetone into the audio output stage, IC3b. The value of R5 sets the sidetone level.

## Constructional notes

The component layout is shown in Fig 5.13, and the components list is provided in Table 5.1. For the construction of the unit, a few precautions should be observed.

1. Check that all top-side solder connections are made.
2. Wind turns onto toroids tightly and fix to PCB with a spot of glue.
3. Do not use IC sockets. Observe anti-static handling precautions for all ICs and FETs.
4. All VFO components should be earthed close to VFO.
5. Fit a 1A fuse in the supply line.
6. Component suppliers: Bonex, Cirkit, Farnell Electronic Components, JAB Electronic Components etc.

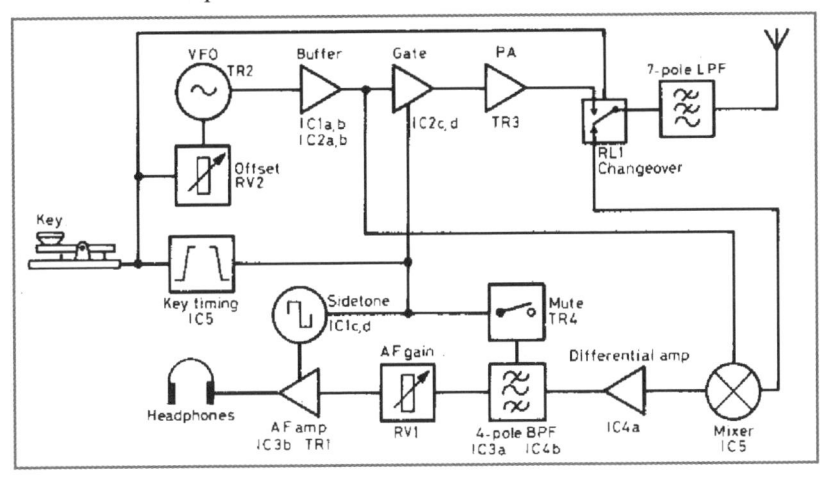

Fig 5.11: Block diagram of the G4ENA transceiver

67

# HF AMATEUR RADIO

Fig 5.12: Circuit diagram of the G4ENA transceiver

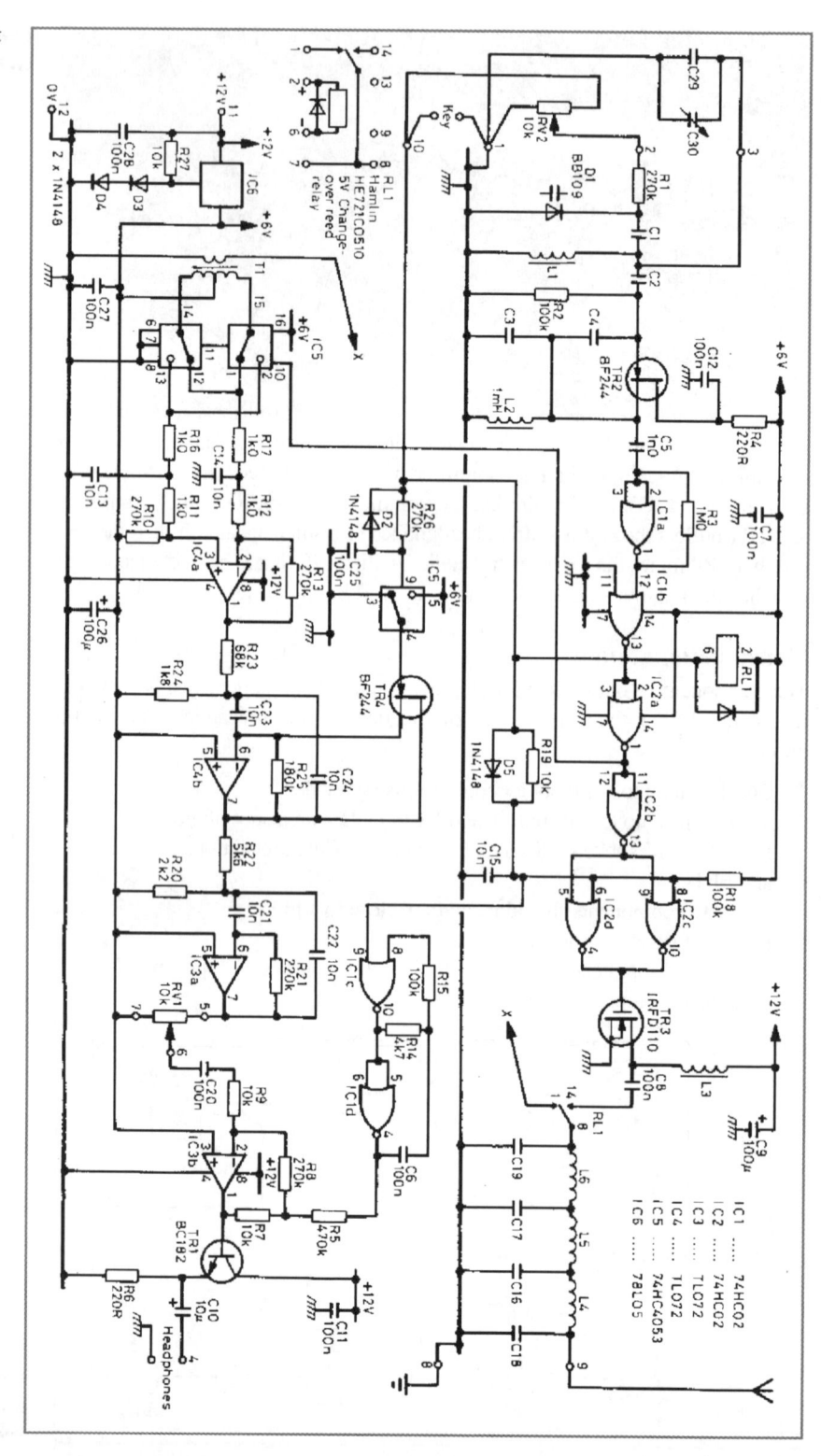

CHAPTER 5: TRANSMITTERS

## Test and calibration

Before TR3 is fitted, the transceiver must be fully operational and calibrated. Prior to switch-on, undertake a full visual inspection for unsoldered joints and solder splashes.

Proceed as follows:
1. Connect external components C29/30, RV1/2, headphones and power supply.
2. Switch on power supply. Current is approx 50mA.
3. Check +6V supply, terminal pin 7. Voltage is approx 6.3V.
4. Select values for C29 to bring oscillator frequency to CW portion of band (1.81-1.86MHz / 3.50-3.58MHz). Coverage should be set to fall inside the band limits of 1.810 / 3.500MHz.
5. Connect an antenna or signal generator to terminal pin 9 and monitor the received signal on headphones. Tuning through the signal will test the response of the CW filter which will peak at about 500Hz.
6. Connect key and check operation of sidetone and antenna changeover relay. Sidetone level can be changed by selecting value of R5.
7. Monitor output of IC2c and IC2d (TR3 gate drive) and check correct operation. A logic low should be present with key-up. and on key-down the VFO frequency will appear. This point can be monitored with an oscilloscope or by listening on a receiver with a short antenna connected to IC2c or IC2d.
8. When all checks are complete fit TR3 (important! - static sensitive device) and connect the transceiver through a power meter to a dummy load. On key-down the output power should be at least 5W for +12V supply, rising to 8W for

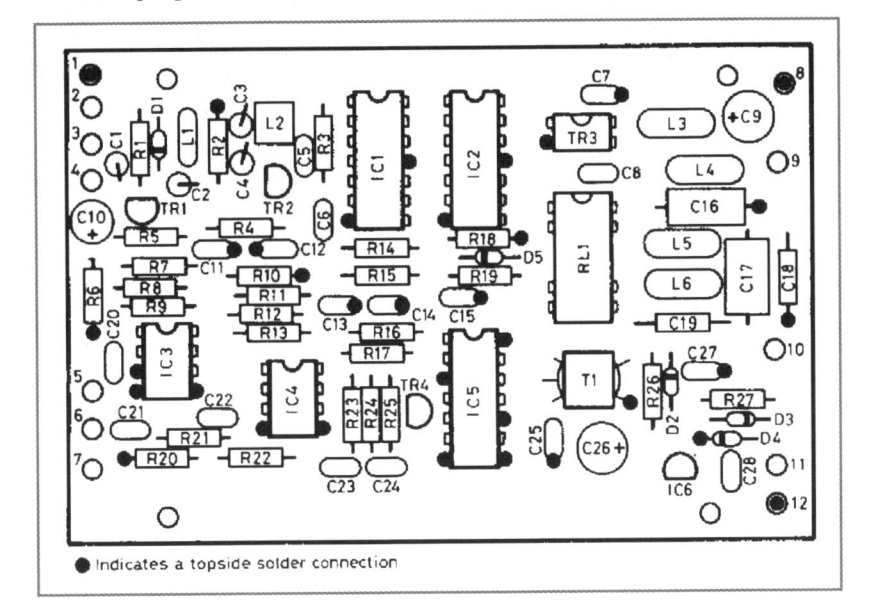

**Photo 5.6:** Inside view of the G4ENA transceiver

**Fig 5.13:** A neat component layout results in a compact unit suitable for portable operation

69

13.8V supply. Note: should the oscillator stop when the key is pressed it will instantly destroy TR3. Switch off power when selecting VFO components.
9. Connect antenna and call CQ. When a station replies note the position of the RIT control. The average receive offset should be used when replying to a CQ call.

## In operation

The QSK (full break-in) concept of the rig is very exciting in use. The side tone is not a pure sine wave but is easily heard if there is an interfering beat note of the same frequency. One important note is to remember to tune the receiver into a station from the high-frequency side so that when replying your signal falls within his passband.

Both the 160 and 80m versions have proved very successful on-air. During the 1990 Low Power Contest, the 80m model was operated into a half-wave dipole and powered from a small NiCad battery pack. This simple arrangement produced the highest 80m single band score!

Its small size and high efficiency makes this rig ideal for portable operation. A 600mAh battery will give several hours of QRP pleasure.

### Table 5.1: Parts list for the G4ENA transceiver

**Resistors**

| | |
|---|---|
| R1, 8, 10, 13, 26 | 270k |
| R2, 15, 18 | 100k |
| R3 | 1M |
| R4, 6 | 220R |
| R5 | 470k |
| R7, 9, 19, 27 | 10k |
| R11, 12, 16, 17 | 1k |
| R14 | 4k7 |
| R20 | 2k2 |
| R21 | 220k |
| R22 | 5k6 |
| R23 | 68k |
| R24 | 1k8 |
| R25 | 180k |
| All resistors 0.25W 2% | |
| RV1, 2 | 10k linear |

## CHAPTER 5: TRANSMITTERS

### Capacitors

| Ref | Type | Pitch | Value(80m) | Value(160m) |
|---|---|---|---|---|
| C1 | Ceramic plate | 2.54 | 4p7 | 15p |
| C2 | Polystyrene | — | 47p | 100p |
| C3, 4 | Polystyrene | — | 220p | 470p |
| C5 | Ceramic monolithic | 2.54 | 1n | 1n |
| C6, 7, 8, 11, 12, 20, 25, 27, 28 | Ceramic monolithic | 2.54 | 100n | 100n |
| C9, 26 | Aluminium radial 16V | 2.5 | 100μ | 100μ |
| C10 | Aluminium radial 16V | 2.0 | 10u | 10u |
| C13, 14, 15, 21, 22, 23, 24 | Ceramic monolithic | 2.54 | 10n 10% | 10n 10% |
| C16, 17 | Polystyrene | — | 1n5 | 2n7 |
| C18, 19 | Polystyrene | — | 470p | 1n |
| C29* | Polystyrene | — | 470p | 820p |
| C30* | Air spaced VFO | — | 25p | 75p |

\* = select on test

### Inductors

| Ref | Type | 80m | 160m |
|---|---|---|---|
| L1 | T37-2 (Amidon) | 31t 27SWG (0.4mm) | 41t 30SWG (0.31mm) |
| L2 | 7BS (Toko) | 1mH | 1mH |
| L3 | T37-2 | 2.2 μH 23t 27SWG | 4.5 μH 33t 30SWG |
| L4, 6 | T37-2 | 2.9 μH 26t 27SWG | 5.45 μH 36t 30SWG |
| L5 | T37-2 | 4.0 μH 31t 27SWG | 6.9 μH 41t 30SWG |
| T1 | Balun | 2t primary, %+5t secondary, 0.2mm 36SWG (28-43002402) | |

### Semiconductors

| | |
|---|---|
| D1 | BB109 |
| D2, 3, 4, 5 | 1N4148 |
| TR1 | BC182 (not "L") |
| TR2, 4 | BF244 |
| TR3 | IRFD110** |
| IC1, 2 | 74HC02** |
| IC3, 4 | TL072** |
| IC5 | 74HC4053** |
| IC6 | 78L05 |

\*\* = static sensitive devices

### Miscellaneous

| | |
|---|---|
| RL1 | 5V changeover reed relay, Hamlin HE721C0510; PED / Electrol 17708131551-RA30441051 |
| PCB | Boards and parts are available from JAB Electronic Components, The Industrial Estate, 1180 Aldridge Road, Great Barr, Birmingham B44 8PE, email jabdog@blueyonder.co.uk. |

CHAPTER 6

# Antennas

*In this chapter:*

- Feeders
- Connectors
- Standing wave ratios
- Antenna tuning units
- Directivity
- End-fed wire
- Dipole
- Multiband antennas
- Vertical
- Beam (Yagi)
- Earth connections
- Selecting a location
- Safety

THE performance of an antenna greatly affects the performance of the station, so time and energy invested in it will pay dividends when operating on the bands. A poor antenna will limit the performance of the station, however good the transmitter, receiver or transceiver may be. Conversely, a good one will make the most of your equipment.

Unfortunately, it is not usually possible to have best antenna system for the station. The garden may be too small or there could be other restrictions that prevent the ideal antennas being installed. But all is not lost if it is not possible to deploy a large system. Often disadvantages in one area can be made up in others, and many people enjoy experimenting with antennas to find the best one for their particular location.

## Constituents of an antenna system

An antenna system is made up of a number of parts, not just the antenna itself. A feeder is required to enable the power to be transferred to the antenna from the station or vice versa. There may also be some matching circuitry between the feeder and the antenna – this is required because the maximum amount of power is transferred from the feeder to the antenna when the impedance of the two is the same.

## Feeder

Feeders (transmission lines) are used to transfer RF energy from one point to another, and they should introduce as little loss as possible. They transfer the

# HF AMATEUR RADIO

signals by propagating a radio frequency wave along their length while not allowing it to radiate. This is done by confining the electric and magnetic fields associated with the wave to the vicinity of the feeder.

There are several forms of feeder, but the two most common for HF applications are coax (or coaxial cable) and the various forms of twin feeder.

## Coax

Coax cable consists of two concentric conductors as shown in Fig 6.1. The centre conductor is almost universally made of copper. Sometimes it may be a single conductor whilst at other times it may consist of several strands.

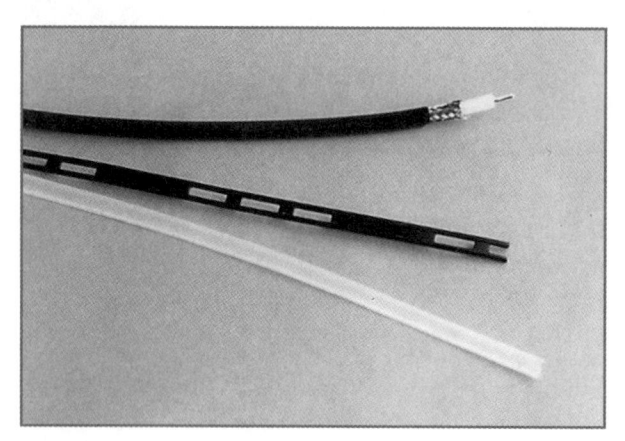

**Photo 6.1: Coaxial cable, black twin and ordinary twin feeder**

The outer conductor is normally made from a copper braid. This enables the cable to be flexible which would not be the case if the outer conductor was solid. To improve the screening, double or even triple screened cables are sometimes used. Normally, this is accomplished by placing one braid directly over another although in some instances a copper foil or tape outer may be used. By using additional layers of screening, the levels of stray pick-up and radiation are considerably reduced. More importantly for most radio amateurs, lower levels of loss are achieved.

Between the two conductors there is an insulating dielectric. This holds the two conductors apart and in an ideal world would not introduce any loss. This dielectric may be solid or, as in the case of many low-loss cables, it may be semi-air-spaced because it is the dielectric that introduces most of the loss. The semi-air space may be in the form of long 'tubes' in the dielectric, or a 'foam' construction where air forms a major part of the material.

Finally, there is the outer cover or sheath. This serves little electrical function, but can prevent earth loops forming. It also gives vital protection against dirt and moisture attacking the cable. However, when burying cable, it is best not to rely on the sheath. Instead, use conduit or special 'bury direct' cables.

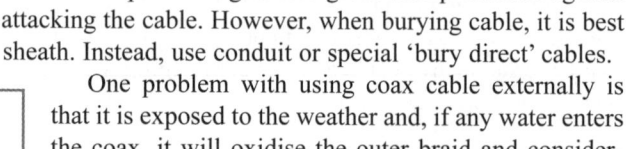

**Fig 6.1: Coaxial feeder**

One problem with using coax cable externally is that it is exposed to the weather and, if any water enters the coax, it will oxidise the outer braid and considerably reduce its conductivity, thus increasing its loss. Additionally, any moisture entering the dielectric spacer will absorb power. This means that every time coaxial cable is used outside, it is necessary to give thought to how to stop water getting in.

One option is to seal the ends with a sealant such as those used for bathrooms. However, the best solution is to use self amalgamating tape. This is made by a number of companies and is available from antenna and general amateur radio stockists. The self-amalgamating

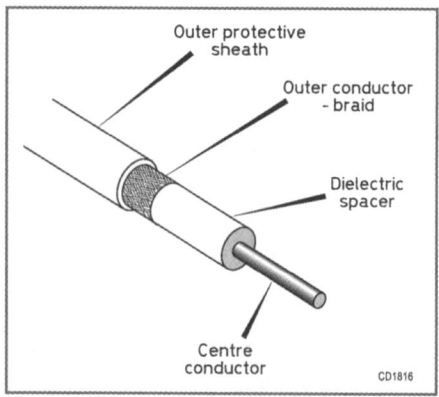

74

tape comes in the form of a roll and appears like thick insulating tape but with a thin paper backing on one side. It is used in a similar way to insulating tape. The backing strip is peeled off and then it is wrapped around whatever it is to be waterproofed. You should overlap each winding by about 50% of its width to ensure a good seal. When applying the tape, keep it stretched so that it goes on under tension. Also, it is best to start from the thinner end of the job, i.e. where the diameter of whatever it is being applied to is smallest. Where there is a connector on a cable, start on the cable and work towards the connector. Also when winding, ensure that there are no holes of voids in which water could condense or enter.

## Twin feeder

Another type of feeder that is sometimes used on frequencies below 30MHz consists of two parallel wires. As the currents in the two wires are equal and opposite, in theory no signal should be radiated from them. In practice, they are affected by nearby objects and this form of feeder should not be run through a house for example.

There are a number of types of twin feeder. The most common consists of two wires spaced about 20mm apart covered in a translucent plastic and is called twin feeder or ribbon feeder. Whilst it is fine for internal use, when used externally the plastic dielectric can absorb water, causing losses to rise significantly. An alternative version using black plastic with ovals cut in the spacing dielectric works well for external use.

It is possible to make open-wire feeder. This can be achieved by taking two lengths of wire, separated at intervals by spacers. These can be made from plastic waste piping (the white variety), which is cut into lengths of about 15cm. This is then cut into strips with holes drilled at either end and attached to the wire as shown in Fig 6.3. The spacers can be placed every metre or so. Obviously, the more spacers that are used, the better the control of the spacing. This type of open-wire feeder can be made very easily and when used away from other objects gives very low levels of loss. Its impedance will typically be around 600 ohms. This feeder is suitable for feeding an antenna when combined with an antenna tuning unit. Any imperfections in the match can easily be tuned out.

The loss that a feeder introduces between the antenna and receiver or transmitter is of the utmost importance, as it will reduce the efficiency of the station. This loss occurs in two main ways. The first is as power radiated from the feeder. Although normally small, this radiated power can become significant especially over long runs. The second is as power being dissipated as heat as a result of the resistance in the feeder. These 'ohmic' losses can be reduced by making the feeder conductors larger so that their resistance is lower.

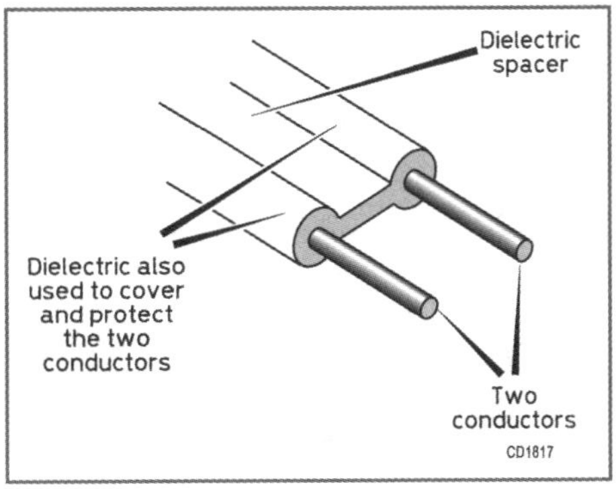

The dielectric between the conductors can also give rise to losses. Power may be dissipated here, especially if moisture is present. It is therefore impor-

**Fig 6.2: Twin or ribbon feeder**

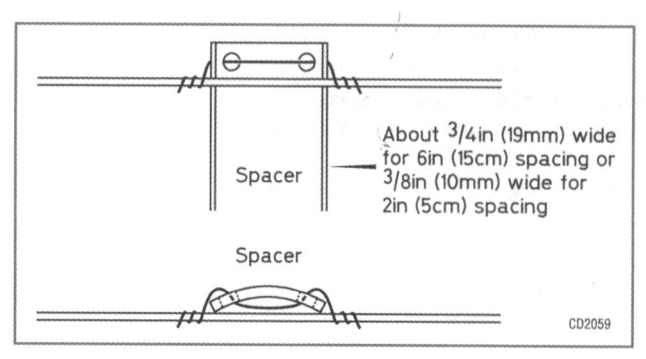

**Fig 6.3: An example of simple and inexpensive construction of 600-ohm open-wire feeder.** Plastic piping of 5 to 8cm diameter is cut into short lengths, and then sawn lengthwise into strips, a pair of holes being drilled near the end of each strip. Spacing may be 5 to 15cm with insulation at intervals of some 12 to 15 times the line spacing. Losses are reduced with wide spacing and construction is much quicker, but symmetry is more easily upset. Note that the spacers are slightly curved and this helps to prevent slippage.

**Photo 6.2: Connectors from left to right, PL259 + reducer, Type N, BNC.**

tant to ensure that moisture does not enter coaxial cable.

## Characteristic impedance and velocity factor

It has already been mentioned that the maximum power transfer from a transmitter to an antenna occurs when their impedances are both the same. Feeders also have an impedance and this is of great importance when designing and erecting antennas.

The impedance of a feeder can be best demonstrated by taking the example of an infinitely long line with no losses. A signal applied to this line would propagate along it forever and never be reflected or returned. As the energy propagating along the wire is always travelling and not stored, the line would look like a pure resistor to the transmitter.

If the line was cut at a finite distance from the transmitter and the end left either open-circuit or short-circuited, then any signal travelling along the feeder would not be able to travel any further. In this case, the only way for it to travel is to be reflected back the way it came.

Now if a variable resistor with its value set higher than the characteristic impedance of the line is connected to the end of the line, some power will be dissipated in this resistor. As the value of the resistor is reduced, more power is transferred to it and less is reflected. Eventually, a point is reached where all the power is dissipated in the resistor and none is reflected. If the value of the resistor is reduced beyond this point, less power is transferred to it and more becomes reflected.

The value of the resistor when no power is reflected represents the impedance of the feeder itself and it is known as the characteristic impedance. Generally, coaxial cables have an impedance of 75 ohms if they are used for domestic television, or 50 ohms if they are for professional, amateur radio or CB use. Other impedances are often used for computer applications so beware of type of cable.

## Connectors

When using coaxial cable, it is also necessary to consider the connectors. Choosing the right connector is important. For amateur radio installations, there are three main types of connector that are used:

- UHF Connector
- BNC Connector
- N-Type connector

Each of these connectors has its own advantages and disadvantages as far as amateur radio operation is concerned. And a short description of each type is merited here.

## UHF connector

The UHF connector is sometimes known as the Amphenol coaxial connector, It was by the Amphenol company for use in the radio industry. The plug may also be referred to as a PL259 coaxial connector, and the socket as an SO239 connector. These names come from military part numbers.

The connectors have a threaded coupling, and this prevents them from being removed accidentally. It also allows them to be tightened to achieve a good low resistance connection between the two halves.

PL259 connectors come in two sizes for thick and thin coaxial cable. Thin coaxial cables are often used for short runs or 'patch' leads but not for long runs as the thinner cables have a higher loss than thick ones. When needed, PL259 plugs are commonly used with a "reducer" to fit the large cable entry hole in the plug to the thin cable.

The drawback of UHF or Amphenol connectors is that they have non-constant impedance. This limits their use to frequencies of up to 300MHz, but despite this these UHF connectors provide a low cost connector suitable for many applications.

Soldering PL259 connectors is not always easy,. The tools required for soldering include a tool to strip the cable, a soldering iron, and solder. A pair of pliers to hold the item being worked on is helpful, especially as the connectors can retain their heat for some while. A vice can also be very useful.

Start by stripping back about 1.5 inches (35mm) of the outer coating or sheath of the cable, taking care not to cut too deeply and score any of the fibres of the conductive braid. Leave around 0.5 inch (13mm) of the copper braid or shielding in place and then remove about 0.5 inch (13mm) of the plastic core.

Tin the exposed central copper core. To do this, heat the core with the soldering iron and apply a thin even coating of solder to it. Take care not to keep the soldering iron on the conductor for too long otherwise the dielectric spacing between the outer and inner conductors of the coax will melt. Once the cable has cooled, slide the inner part of the PL259 plug over the cable with a screwing action until the copper core appears at the end of the centre pin. The trimmed shield will have become trapped between the core and the inside of the PL259. The outer sheath or covering or covering of the coax cable will ensure a snug fit and any protruding shielding should be removed with the sharp knife.

Photo 6.3: Cable trimmed ready for the PL259 plug along side finished example.

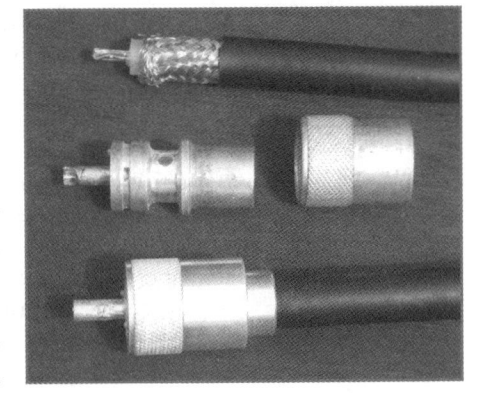

With the soldering iron, heat the centre pin of the PL259 and cable core. Add solder to fill the void in between the core and the plug. Once cool, trim off any protruding core and screw back on the outer cover of the PL259. The plug is now ready for use.

## BNC connector

The BNC coax connector is widely used in professional circles, notably in oscilloscopes and other laboratory instruments. The BNC connector is also widely used when RF connections need to be made. The BNC connector has a bayonet fixing to prevent accidental disconnection while being easy to disconnect when necessary.

Electrically, the BNC coax cable connector is designed to present a constant impedance and it is most common in its 50 ohm version, although 75 ohm ones can be obtained. It is recommended for operation at frequencies up to 4GHz and special top quality versions can be used for up to 10GHz. The BNC connector is not normally used on HF rigs, its use being reserved more for operation in the VHF and UHF regions.

## N-type connector

The N-type connector is a high performance RF coaxial connector used in many RF applications. The connector has a threaded coupling interface to ensure that it mates correctly. It is available in either 50 ohm or 75 ohm versions. These two versions have subtle mechanical differences that do not allow the two types to mate. The connector is able to withstand relatively high powers when compared to the BNC or TNC connectors. The standard versions are specified for operation up to 11GHz, although precision versions are available for operation to 18GHz.

The N-type coaxial connector is used for many radio frequency applications including broadcast and communications equipment where its power handling capability enables it to be used for medium power transmitters. Within amateur radio, its use tends to be restricted to more exacting applications where high power and high frequencies are used.

## Fixing connectors to cables

Connectors are not always easy to fit to cables but it is essential that you get it right. The main secret to success is to use the right cable with the right connector. Not only is it necessary to choose connectors for the correct impedance, but also the correct cable. The different variants of each connector are designed to be used with a particular type of cable. There are also crimp and solder versions available.

Cables are commonly of one of two families: the American 'RG' (Radio Guide MIL specification) types and the British 'UR' (UniRadio) series. URM67 is equivalent to RG213, is 10.5mm diameter and is the most common feeder used with N-type and PL259 connectors. URM43 (5mm OD) is usually used with BNC connectors, although it also fits RG58 cable since both have similar dimensions.

The connectors usually contain a number of different parts, and these can be seen in Fig 6.4. All original clamp types use a free centre pin held in place by its solder joint on to the inner conductor. Captive contact types have a two-part centre insulator between which fits the shoulder on the centre pin. Improved MIL clamp types may have either free or captive contacts. Pressure-sleeve types have a captive centre pin. Pressure clamp captive pin types can be identified by the fact that they have a ferrule or 'top hat' that assists in terminating the braid, a two-piece insulator and a centre pin with a shoulder. Unimproved clamp types have a washer, a plain gasket, a cone-ended braid clamp and a single insulator, often fixing inside the body. Improved types have a washer, a thin ring gasket with a V-groove and usually a conical braid clamp with more of a shoulder.

Further details of how to fit connectors to cables can be found in Chapter 14 of the Radio Communications Handbook 8th Edition – see further reading.

## Standing wave ratio

When a load (or in this case an antenna) is not properly matched to the feeder, then not all the power reaching the end of the latter can be transferred to the load. However, the power cannot just disappear and has to go somewhere. The only place that it can go is back along the feeder, which gives rise to a number of effects. High voltage and current points are set up along the feeder and, especially when high-power transmitters are used, these may cause damage to equipment. These points may also appear at the output of the transmitter and may cause damage to the output devices if no protection circuitry is present. If this circuitry is present, then the high level of standing waves will be detected and the transmitter power may be reduced. Finally, the high levels of standing wave may mean that the antenna system is not operating efficiently.

The level of standing waves is normally quoted as a standing wave ratio (SWR). A ratio of 1:1 shows that there is a perfect match and no power is being reflected. A ratio of infinity to 1 shows that there is a complete mismatch and all the power is being reflected. A ratio of 2:1 is normally acceptable although it is best to aim for the minimum attainable.

## Antenna tuning unit

An antenna tuning (matching) unit (ATU) is used to give a good match between the feeder and antenna. These are widely available, consisting of vari-

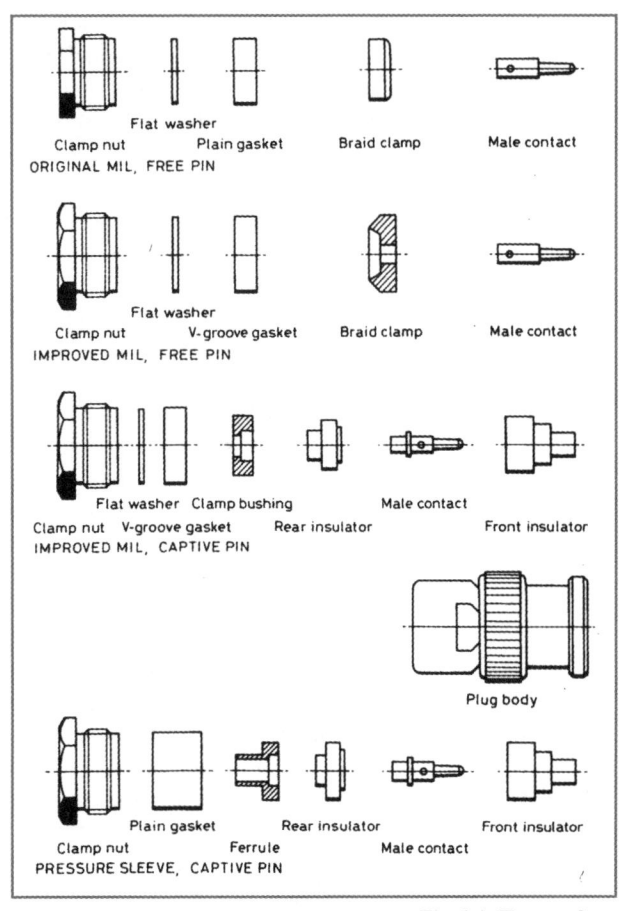

**Fig 6.4: Types of BNC and N cable clamps.**

# HF AMATEUR RADIO

able capacitors and inductors, and are an essential part of many antenna systems. Ideally, the ATU is placed at the interface between the feeder and the antenna itself, but this is often not practical. In some antenna systems, such as the open-wire-fed doublet or the end-fed wire, the ATU may be placed between the coaxial feed from the transmitter and the end of the open-wire feeder or wire antenna. Where coaxial feeder is used all the way from the transmitter to the antenna, an ATU located close to the transmitter can be used to reduce the SWR seen by the transmitter's power amplifier. In this way, it may enable the transmitter to reach its full output as many sets include a system that will reduce the power output under conditions of high SWR to protect the output devices.

## Directivity

An important aspect of any antenna is the way in which it radiates in different directions. No antenna radiates equally in all directions, and as a result the radiation pattern around it is of interest – a plot of this pattern is known as a polar diagram.

Diagrams for two types of antenna are shown in Fig 6.5. The first example is that of a half-wave dipole. This has a figure-of-eight shape, radiating most of the signal at right-angles to the axis of the antenna and least along its length. Other antennas have more pronounced radiation patterns. Some antenna are designed specifically to 'beam' power in one particular direction and are therefore called beams. Such antennas are also very good at receiving signals from one particular direction, and are often used for DXing.

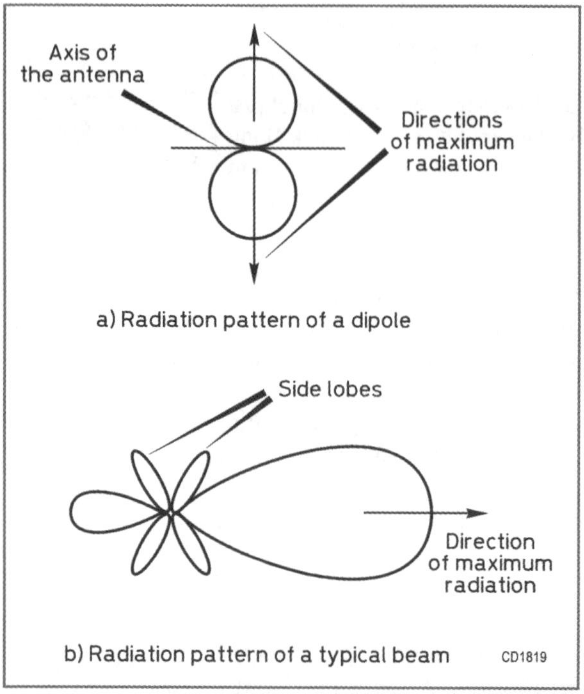

As more power is radiated in some directions, the antenna is said to have gain in the direction of its maximum radiation. It should be remembered that antennas with gain will radiate less in other directions, which can be used to advantage. The transmitted signal can be concentrated only in the direction where it is needed, reducing the level of interference to stations in other directions. Conversely, when receiving, the level of interference caused by stations that are not on the direction of the beam is reduced. This can be a considerable advantage when the band is crowded with strong stations.

The gain of the antenna is expressed in decibels and is usually compared to a dipole antenna where the gain is measured as dBd (decibels relative to a dipole). In some instances, the gain will be compared to an isotropic source (a theoretical antenna that radiates equally in all directions) and expressed as dBi.

Fig 6.5: Radiation patterns of two types of antenna

Beware when comparing gain for different antennas because figures compared to an isotropic source are 2.1dB higher than those compared to a dipole and either reference may be used in a specification.

Often, equally important as the forward gain is the front-to-back ratio. This figure is the ratio between the power in the forward direction of the antenna to the power in the reverse direction and this is expressed in decibels. When making adjustments to an antenna, it is found that the optimum forward gain and front-to-back ratio do not usually coincide. In many instances people will choose the maximum forward gain but, when high levels of interference are normally experienced in the direction opposite to the wanted stations, the maximum front-to-back ratio may be the best option.

It should be remembered that when a directional antenna is used, it should have some means of orientating it in the required direction. Most stations using a beam will use a rotator, which is a unit that has a motor on the mast head to physically rotate the beam, and a controller that can be located by the equipment. This controller remotely controls the direction of the antenna, allowing it to be orientated in the required direction. Given the need for the rotator and the associated mast, such as system is considerably more expensive than an antenna alone.

## End-fed wire

One of the easiest forms of antenna to erect is an end-fed wire. It simply consists of a wire from the transmitter and receiver strung between two high points. The wire is often cut to a quarter- or a multiple of quarter of the wavelength of the frequency being received.

A wire of this nature is very easy to install, requiring little in the way of special feeders. If it is only to be used for receiving, then it is possible to connect it just to the input of the receiver, although for the best performance an antenna tuning unit is required. This will also be essential if the antenna is to be used for transmitting.

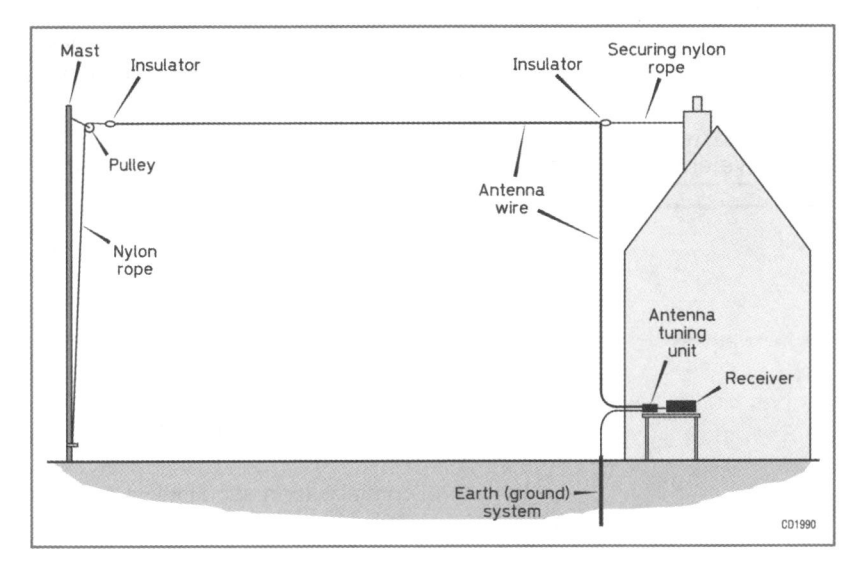

Fig 6.6: A typical end-fed wire installation

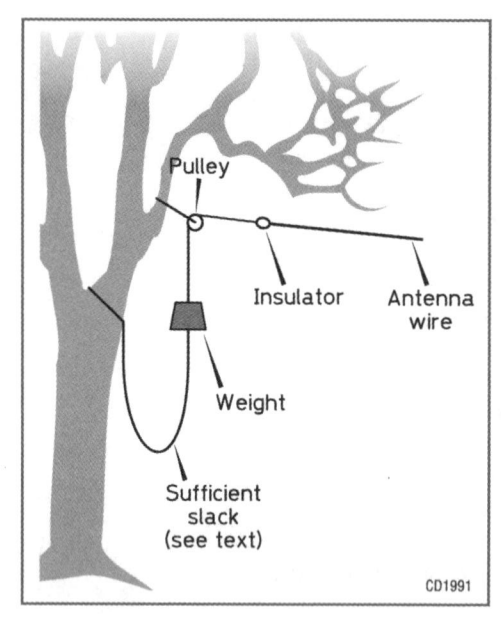

# HF AMATEUR RADIO

**Fig 6.7: Anchoring a wire antenna to a tree**

**Fig 6.8: A half-wave dipole**

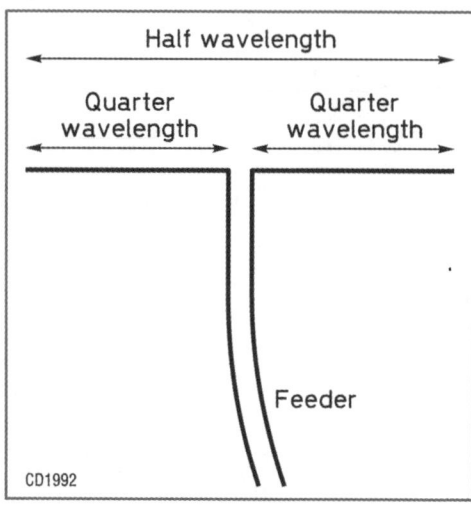

A typical installation is shown in Fig 6.6. In this instance, the wire leaves the antenna tuning unit and passes out of the house.

With the installation shown, the vertical section travels up to meet the horizontal section. Nylon rope can be attached to either end, and insulators used. These can be bought from amateur radio stockists, and provide a neat way of terminating the wires as shown. Proper fixings should be used to attach the rope to the house so that it does not wear against the brickwork.

It is often convenient to use a tree as one of the anchor points for the antenna. If this is done, then provision must be made to accommodate any movement due to the wind. This can be achieved by using a system like that shown in Fig 6.7. Here, a constant tension is held on the wire regardless of the movement of the tree. However, there should be sufficient rope in the loop to accommodate the movement under the windiest conditions, and care should be taken to ensure that the loop will not become snagged in the tree.

It must be said that this type of antenna is far from ideal. It radiates along its whole length and this means that there are likely to be high levels of radio frequency radiation in the shack that could be a health hazard, particularly if high power levels are used. It can also cause interference to household electronics or to other pieces of equipment in the shack – sometimes RF can be picked up by the microphone input to the transmitter, causing the audio to become distorted.

For an end-fed wire to operate satisfactorily, it should have a good earth system. In many instances, this is given less attention and the effectiveness of the whole antenna is reduced.

Often the name 'long wire' is used for these antennas. Although the term normally applies to an ordinary end-fed wire, it strictly refers to an antenna that is several wavelengths long. An antenna like this is highly directional, being what is called an end-fire antenna. As the name suggests, the direction of maximum radiation is along the axis of the antenna itself.

## Dipole

The dipole is one of the most important types of antenna. It is widely used in its own right, and also forms the basis of a number of other antennas that are very popular.

In its most common form, the dipole is an electrical half-wavelength long, and fed in the middle as shown in Fig 6.8. Although it is usually thought of as a half-wave antenna, it can be made any number of half-wavelengths. This means that an antenna that is cut to operate as a half-wavelength dipole on one band will also operate as a three-half-wavelength dipole on a band that is at three times the frequency. For example, a half-wavelength antenna for 7MHz can be used as a three-half-wavelength antenna on 21MHz.

The polar diagram of an antenna varies according to its length in wavelengths. A half-wave dipole has the typical figure-of-eight polar diagram but as the length is increased it develops lobes which tend to align progressively with the axis of the antenna. This can be seen from Fig 6.9.

The impedance of the antenna is important. A dipole operating in free space has a feed impedance of 73 ohms but nearby objects, including the ground, will alter this, and 50 ohm cable will usually give a good match.

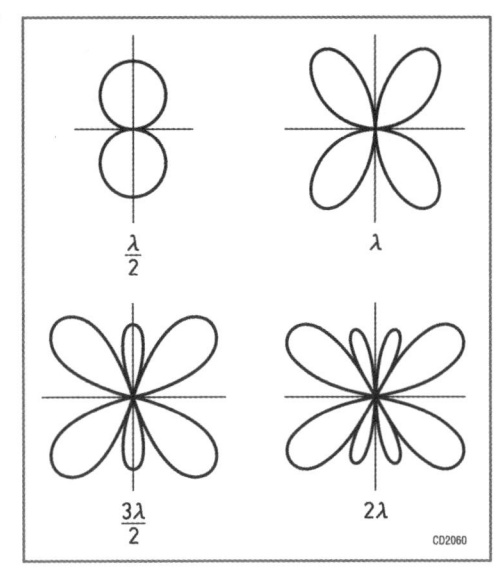

Fig 6.9: Polar diagrams for half-wavelength and three-half-wavelength antennas

The length of the antenna is quite critical if it is to work properly – it must be an electrical half-wavelength (or multiple of half-wave-lengths). This length is not quite the same as the wavelength in free space. There are a number of reasons for this; one is called the end effect and is due to the fact that the antenna is not infinite. The length of a half-wave dipole can be calculated from the formula:

$$\text{Length(m)} = \frac{148}{\text{Frequency(MHz)}}$$

$$\text{Length(ft)} = \frac{480}{\text{Frequency(MHz)}}$$

Using the formula, it can be seen that lengths for antennas in the HF bands are as shown in Table 6.1. However, it should be remembered that these are approximate lengths and the antenna should be cut slightly longer first and trimmed to give the optimum operation.

## Building a dipole

The most straightforward way to install a dipole is as a horizontal antenna, although this is by no means the only way and various other methods and types will be described later. The basic dipole is shown in Fig 6.10. As with the end-fed wire, some means of strain relief is required to account for any movement if a tree is used as one of the anchor points. Again,

Table 6.1: Half Wave Dipole Lengths for the HF Amateur bands

| Band (MHz) | Length (feet) | Length (metres) |
| --- | --- | --- |
| 1.8 MHz | 266 | 82.2 |
| 3.5 MHz | 137 | 42.2 |
| 7.0 MHz | 68.5 | 21.1 |
| 10.1 MHz | 47.5 | 14.7 |
| 14.00 | 34.3 | 10.6 |
| 18.068 | 26.6 | 8.2 |
| 21.00 | 22.8 | 7.04 |
| 24.89 | 19.3 | 5.94 |
| 28.00 | 17.1 | 5.28 |

insulators should be used at either end and nylon rope can be used to secure the dipole itself.

The centre of the dipole requires the coaxial or open-wire feeder to be connected to it. While it may be tempting to simply connect the feeder and let it take the strain, this is not particularly satisfactory when there is a long drop for the feeder – a dipole centre should be used. This will take the strain caused by the tension on the wire, thereby avoiding damage to the feeder over a period of time. Often it is possible to use an ordinary antenna insulator for this purpose.

While it may not always be possible, it is certainly helpful to route the feeder away from the antenna at right-angles to the antenna wire as shown. In this way, the feeder will have the minimum effect on the operation of the antenna.

Often a balun is placed at the feed point of the dipole. This is a transformer used to connect a balanced system to an unbalanced one, or vice versa. It is required because a dipole is a balanced antenna, i.e. neither connection is earthed, and coaxial feeder is unbalanced, having the outer braid of the feeder connected to earth. Although the antenna will operate without a balun, the use of one will prevent signals being radiated from or picked up by the braid on

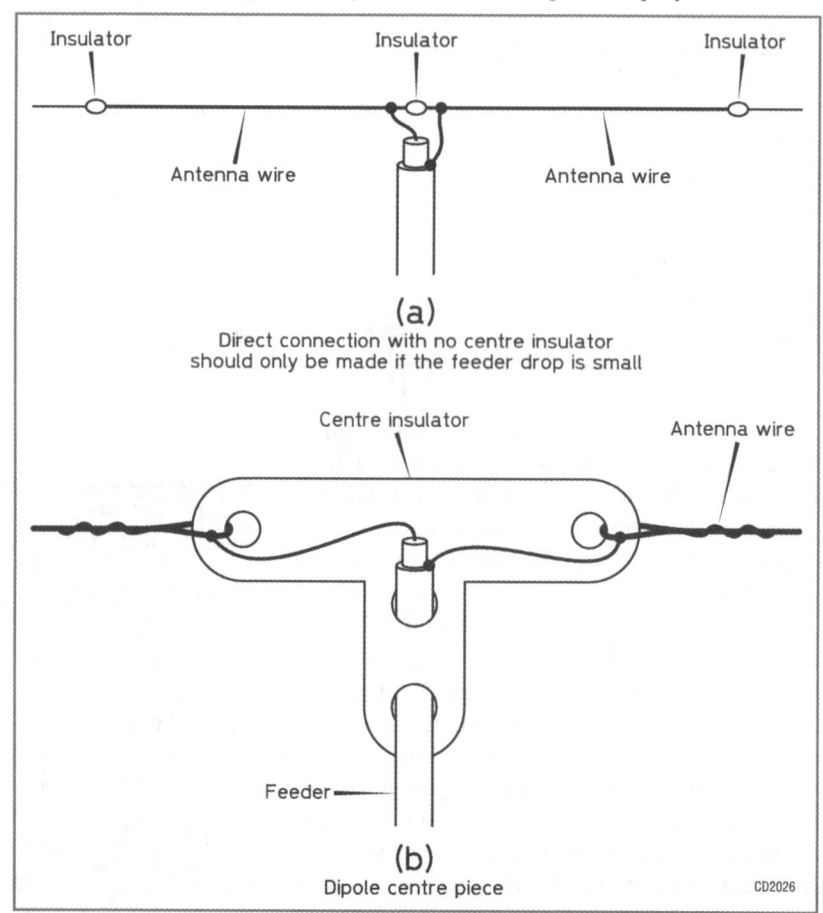

Fig 6.10: Dipole Construction

the feeder. This may help prevent interference to nearby televisions or other radio equipment. The use of a balun also ensures that the normal figure-of-eight radiation pattern is preserved.

Baluns can be made or bought. In the case of feeding a dipole with 50-ohm coax they would normally be a 1:1 transformer, i.e. one having the same number of turns on the primary and secondary.

## Sloper

It may not always be convenient to erect a horizontal dipole. There may be insufficient space or other performance issues. Fortunately, it is possible to use the dipole in a number of alternative ways to enable it to be fitted into the available space.

One configuration that can be used is known as a sloper. It can be used in situations where a single support, preferably a non-metallic mast or a high point on a building, is available. The antenna can be arranged to slope down towards the ground at an angle between 45 and 60° (Fig 6.11). Ideally, the sloping half-wave dipole should have its lower end at least one-sixth of a wavelength above ground, and its feeder should come away from the radiator at 90° for at least a quarter of a wavelength. If coaxial feeder is used, the braid should connect to the lower half of the antenna to help feed balance.

The performance of a sloping dipole is quite different from a horizontal version and it can be good for long distance work. The radiation from a sloping dipole shows slant polarisation with both vertical and horizontal components according to the amount of slope. Its lower angle of radiation to the horizon can result in useful low angle gain over a horizontal dipole. Some claim this gain may lie between 3 and 6dB but others give lower figures. Whatever the exact level of gain, it compares favourably with some of the cheaper multi-element trapped beams, especially on the low bands. Good gain is difficult to realise on the low bands in other ways, and for most amateurs multi-element Yagi beams are out of the question.

There is some high-angled radiation (see Chapter 2 Propagation for a description of angle of radiation) from the sides of the sloping dipole but very little radiation from its high end. A disadvantage of this type of antenna is that long-distance working is only possible in one direction, but this may be overcome by having a group of three or four 'slopers' suspended from a common central support, each with its individual feed line.

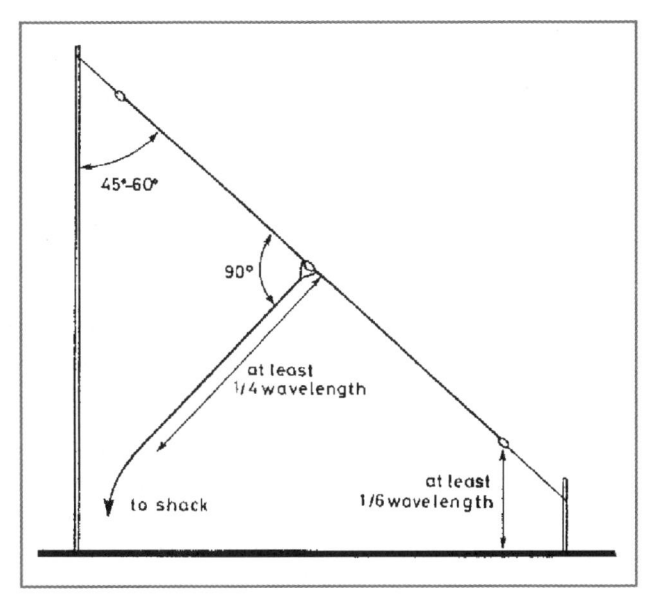

**Fig 6.11:** A half-wave sloping dipole which can be put up in a small space and which will be useful for long-distance working. Most of its low-angle radiation is towards the low end of the antenna but there is also considerable radiation at high angles in other directions. If possible, non-metallic masts should be used to support a 'sloper', but when this cannot be arranged ensure that the mast length is not close to a half-wavelength at the operating frequency.

# HF AMATEUR RADIO

Slopers are ideal in many applications where a single support is available. Many people who have beams and towers mount at least one sloper on the tower for use on the lower frequency bands.

## Inverted-V antenna

One antenna that is particularly convenient is the inverted V dipole. It gains its name from the fact that it has the shape of an inverted V. This type of antenna has a number of advantages. One is that the maximum radiation from any antenna is from the points of high RF current, and a half-wave dipole has this maximum at its centre and for a few feet on either side of the feeder connections. Therefore, it is best to make the centre of the dipole as high as possible. If it is only possible to have one high support, an inverted-V arrangement is obviously ideal. In this way, it is possible to use one fairly high mast in the centre of a garden or plot in locations where the erection of a pair of similar supports with their attendant guy wires would be difficult. A roof-mounted or chimney-mounted mast may also serve as the centre support for a 'V', and the two ends of the dipole can then drop down on either side of a house or bungalow roof. Such chimney mounting will allow the feeder to be dropped to the shack quite easily if it is located in the house.

Although an inverted-V has its greatest degree of radiation at right angles to its axis, its radiation pattern is more omni-directional than that of a horizontal dipole as a result of the fact that the legs are angled downwards.

The inverted-V has an excellent reputation for DX communication on the lower-frequency amateur bands where the installation of large verticals or high horizontal dipoles is not practicable. There are, however, some issues that must be considered when contemplating making one.

The angle between the sloping wires must be at least 90° and preferably 120° or more, as shown in Fig 6.12. This angle dictates the centre support height as well as the length of ground needed to accommodate the antenna. For example, when designed for the 3.5MHz band, an inverted-V will need a centre support at least 14m (45ft) high and a garden length of around 34m (110ft). By contrast, a horizontal dipole needs at least 40m of garden and that neglects to take into account guys to the rear of the end support masts. On the other hand, an inverted-V for operation on 20m (14MHz) only needs a lightweight 5m (15ft) pole to hold up its centre.

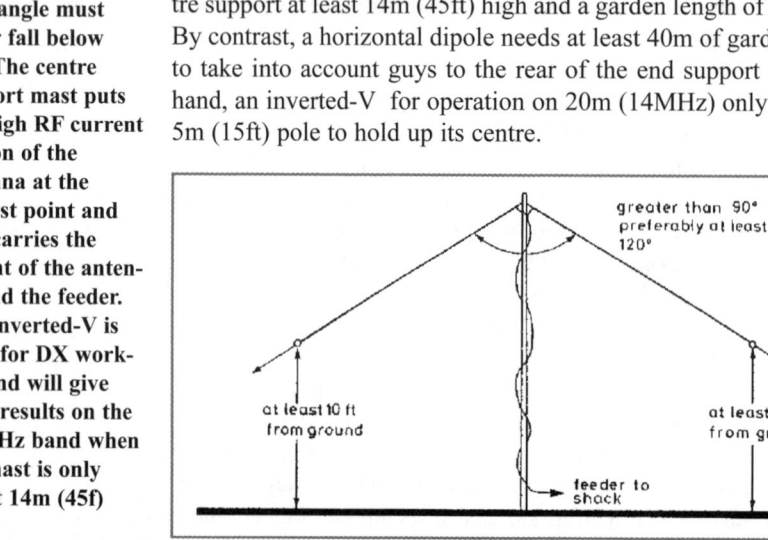

Fig 6.12: A half-wave inverted-V dipole with the angle between the top wires at 120°. This angle must never fall below 90°. The centre support mast puts the high RF current section of the antenna at the highest point and also carries the weight of the antenna and the feeder. The inverted-V is good for DX working and will give good results on the 3.5MHz band when the mast is only about 14m (45f) high.

86

The sloping of the dipole wires causes a reduction of the resonant frequency for a given dipole length, so about 5% must be subtracted from standard dipole dimensions. The calculated wire lengths for inverted-V dipoles on the amateur bands are given in Table 6.2.

A further consequence arising from sloping the dipole wires is a change in the antenna's radiation resistance. The centre feed impedance falls from the nominal 75 ohms of a horizontal dipole to just 50 ohms. This of course is ideal for matching the antenna to standard 50 ohm impedance coaxial cable. An inverted-V antenna has a higher Q than a simple dipole so it tends to have a narrower bandwidth.

**Table 6.2: Suggested lengths for inverted-V dipoles**

| Frequency (kHz) | Length (ft) | (m) |
|---|---|---|
| 3600 | 123' 6" | 37.74 |
| 7050 | 63' 1" | 19.27 |
| 10,100 | 44' 0" | 13.45 |
| 14,200 | 31' 4" | 9.57 |
| 18,100 | 24' 7" | 7.69 |
| 21,200 | 20' 11" | 6.41 |
| 24,940 | 17' 10" | 5.45 |
| 28,200 | 15' 9" | 4.82 |
| 29,200 | 15' 3" | 4.65 |

It is not recommended that the ends of an inverted-V are allowed closer to the ground than about 3m (10ft), even on the higher-frequency bands, because there is the danger of people or animals touching the wire ends. These ends carry high RF potential when energised and could result in a nasty shock or RF burns if touched.

Coaxial feed is recommended for use with an inverted-V, and the low-loss heavier varieties of cable can be used to advantage, for there are no sag problems when the feeder is fastened up at the top and also down the length of the mast. The feeder will impose no strain on the antenna or the soldered connections at its feed point. As with an ordinary horizontal dipole, a balun may be used, although the antenna may operate satisfactorily without one.

## Multiband dipole

One of the disadvantages of a dipole is that it is essentially a single-band antenna, although it can be used on a frequency of three times the fundamental. As most people want to use several bands without having a large array of separate antennas or feeders, there is a need for a more flexible type of antenna that can be used for a range of frequencies.

Remember that a half-wave dipole resonant on one frequency will also resonate at three times this frequency, where it becomes a three half-waves dipole. A 7MHz half-wave dipole for example will also resonate as a three half-wavelength dipole on 21MHz.

Another approach is to use the same feeder for several dipoles. To see how this works, take the example of a dipole cut for 7MHz. This will have the normal low centre feed impedance on that frequency. Should a second dipole which has been cut for 14MHz also be connected to a common feed point, this second and shorter dipole will not present a centre impedance able to accept power at the lower frequency of 7MHz. The converse will apply when the 14MHz dipole is driven, for then the 7MHz dipole will become a centre-fed full-wave antenna and its high centre impedance will not affect the working of the shorter 14MHz wire.

When using this technique, remember that there is no need to include a dipole for a band where one has already been included at a third of its frequen-

# HF AMATEUR RADIO

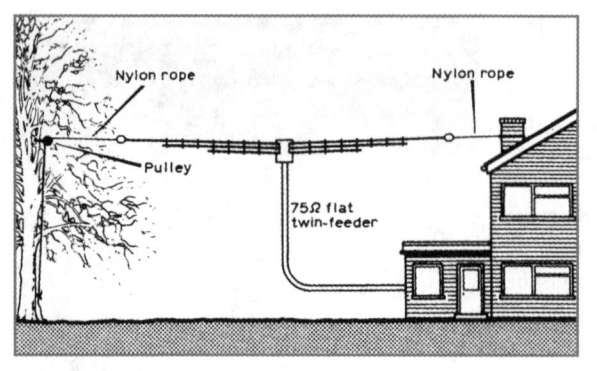

**Fig 6.13: Overall layout of the GI4JTF multiple HF dipole**

cy, i.e. if a 7MHz element is present then there is no need for one at 21MHz.

One design for a multiple dipole system was developed by E Squance, GI4JTF, and appeared in the March 1982 issue of RadCom. The design shows that the lengths that would be expected from the straight dipole length formula are not exactly correct and need to be shortened as a result of the interaction between the different elements of the antenna for the different bands. The lengths are given in Table 6.3.

The antenna is fed with 75 ohm flat twin feeder which is taken back to an ATU. The antenna wire used was multi-stranded 2mm copper wire spaced every 660mm (2ft) using old felt tip pen bodies that were cut and drilled. Overall details of the construction and layout of the antenna are shown in Fig 6.13 and the attachment of the feeder wires and spacers is shown in Fig 6.14.

The centre of the dipole consisted of a polythene plate measuring 110 x 70 x 5mm (4.5 x 2.75 x 0.19in) and is shown in Fig 6.15. Two strips of copper measuring 63 x 12.7mm (2.5 x 0.5inch) were bolted to the plate, one either side to prevent the pull of the wires pulling the bolts through the polythene. The wires were then fed through the centre of the dipole centre plate for rigidity and soldered to copper strip. The feeder was passed through the plate from front to back to prevent whipping when in use. The whole assembly was then liberally coated in marine varnish before being hoisted to the working height.

## Trap dipole

Another popular method of creating a multiband antenna is to isolate sections of it to provide a variety of different resonant lengths. This is achieved by using circuits known as traps. These are parallel-tuned circuits, and at resonance they have a very high impedance. An antenna can be made to resonate on a number of different frequencies by placing one or more of these traps in each leg of it.

A typical example of an antenna using traps is shown in Fig 6.16. Here the centre part of the antenna is resonant on 14MHz because traps resonant at this

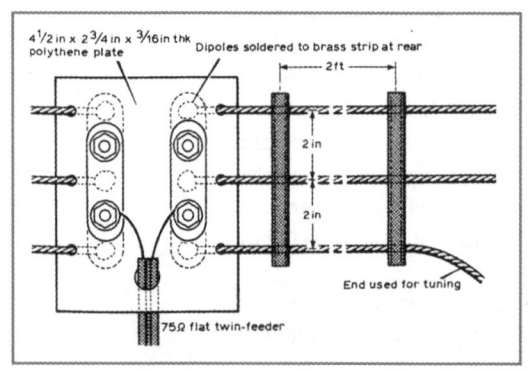

**Left, Fig 6.14: Attachment of the feeder wires and the spacers**

**Below, Fig 6.15: End view of the centre plate showing the mechanical details**

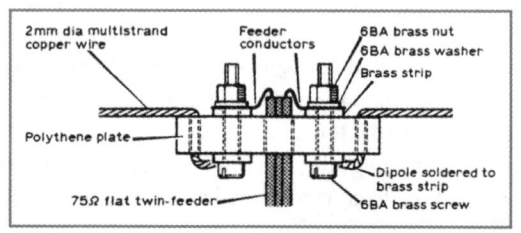

frequency isolate the remainder of it when a 14MHz signal is applied. When the frequency moves away from the resonance of these traps, signals can pass through to use the rest of the antenna. The next resonant length occurs at 7MHz, and again traps resonant at this frequency prevent signals passing any further along the antenna, again isolating the outer sections of the antenna. Finally, the outermost part of the antenna is included and the whole antenna is resonant at 3.5MHz.

**Table 6.3: Lengths for GI4JTF multiple dipole system**

| Resonant Frequency (kHz) | Length of Dipole (m) | (ft) |
|---|---|---|
| 7050 | 20.57 | 67.5 |
| 14,250 | 10.31 | 33.82 |
| 28,775 | 5.08 | 16.67 |

When using a trap dipole, it should be remembered that the physical length of any paths that pass through a trap will be much shorter than their electrical lengths. This is because the trap will appear inductive below its resonant frequency and this inductance will have the effect of shortening the physical length required.

One commercially available version was the G8KW which was fed with 75 ohm coax. Another similar design is shown here (Fig 6.17), although this one is fed with 50 ohm coax and is in an inverted-V format.

Each leg of the antenna consists of two lengths of wire, the one closest to the centre is 32ft 6in long, whereas the outer section is 21ft 6in. The traps comprise 23 turns of 18SWG wire on a 30mm (1.25in) diameter former with a winding length of approximately 65mm (2.5in). The capacitor is 50pF and must be a high voltage type to withstand the antenna end voltage when used with a transmitter. Before connecting the trap to the antenna, it is made resonant at 7.1MHz.

The antenna forms a half-wave on 80m. On 40m, it uses the trap and similarly is a half-wave antenna. On the higher frequency bands, it acts as a multiple half-wave antenna, and as expected there are sharp lobes. Nevertheless, it makes a very useful multi-band dipole.

## Doublet antenna

This antenna is a form of dipole but, because it is fed with open-wire feeder, it is possible to use an antenna tuning matching unit at the end of the feeder to bring the whole antenna to resonance. As a result of this arrangement, these antennas are often called tuned feeder antennas.

When designing or building one of these antennas, the length of the dipole element should be at least a half-wavelength on the lowest frequency of oper-

Below left, Fig 6.16: A trap dipole. The traps isolate sections at different frequencies, giving different effective lengths and allowing multi-band operation.
Below right, Fig 6.17: The inverted-V trap dipole.

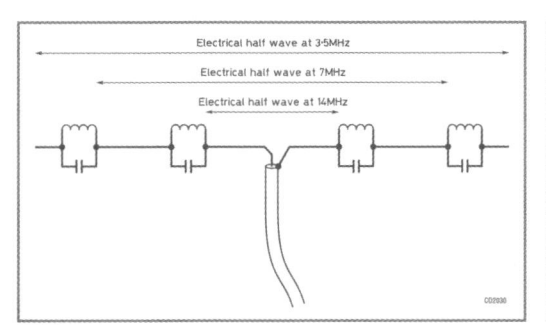

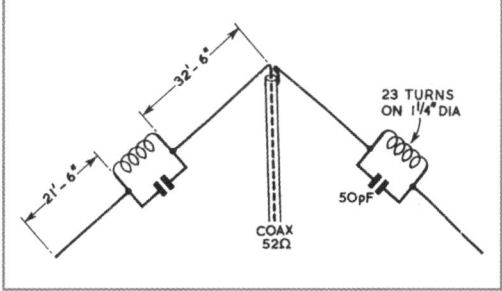

# HF AMATEUR RADIO

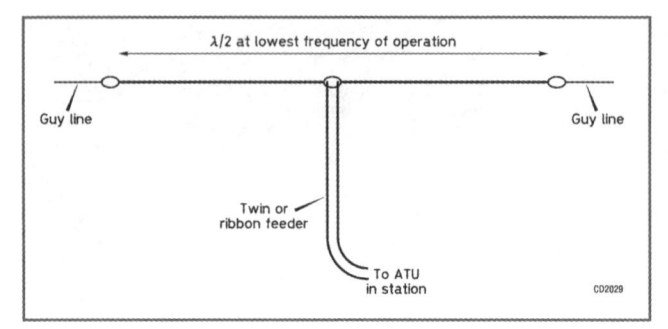

Fig 6.18: A doublet antenna serves as a versatile multiband antenna. The ATU that is used must have a connection for a balanced feeder

ation, but apart from this there are few restrictions. Changing from one band to the next simply requires the antenna tuning matching unit to be readjusted to give the optimum VSWR. Once this has been undertaken a few times the settings can be noted for a much speedier change for the future.

The main drawbacks of this type of antenna are that the tuning unit requires adjusting for each band change; the open-wire feeder should not be run through the house or close to any objects that might unbalance it; and the tuner must have a balanced output. Not all tuning units have this type of output so it is worth checking before buying a new tuner.

The advantage of this type of antenna is that it can be used for all bands above the lowest frequency. This lowest frequency is determined by the length of the element corresponding to a half-wavelength at that frequency. For example, one that is 132ft (40m) long could be used on all the amateur bands from 80m through to 10m.

## G5RV Antenna

Another design for a tuned feeder antenna is the famous G5RV antenna. This was originally designed by Louis Varney, G5RV, in 1946, and some updates were published later in the July 1984 edition of Radio Communication. It is one of the most famous multi-band antennas with countless versions made either by individuals or commercially manufactured. The basic antenna can be seen in Fig 6.19. The G5RV consists of a top radiating section of two 15.6m (51ft) legs. A section of open wire feeder 10.36m (34ft) long is then used to provide a match to either 75 ohm coaxial or twin lead feeder. An ATU may also be substituted at the bottom of the open wire feeder to give a good match to the coaxial feeder. The ATU is needed because the match at the bottom of the open wire feeder is not always optimum. Even if an ATU is not used at this point, one should be used close to the transmitter to enable it to provide a good match at all times.

The antenna can be used on all the HF bands, performing in a number of different modes. On 1.8MHz, the two feeder wires at the transmitter end are connected together, or the inner and outer of the coaxial cable are joined and the top plus 'feeder' used as a "Marconi" antenna with a series-tuned coupling circuit and a good earth connection.

On the 3.5MHz band, the electrical centre of the antenna commences about 5m (15ft) down the open line (in other words, the middle 10m (30ft) of the dipole is folded up). The antenna functions as two half-waves in phase on 7MHz with a portion 'folded' at the centre. On these bands, the termination is highly reactive and the ATU must of course take care of this if the antenna is to load satisfactorily and radiate effectively.

## CHAPTER 6: ANTENNAS

At 14MHz, the antenna functions as a three half-wavelength antenna. Since the impedance at the centre is about 100 ohms, a satisfactory match to the 75 ohm feeder is obtained via the 10.36m (34ft) of half-wave stub. By making the height a half-wave or a full-wave above ground at 14MHz and then raising and lowering the antenna slightly, an excellent impedance match may be obtained on this band. If, however, low-angle radiation is required, height is all important and, as most cables will withstand an SWR of 2 or greater, any temptation to improve the SWR by lowering the antenna should be resisted.

On 21MHz, the antenna works as a slightly extended two-wavelength system or two full-waves in phase, and is capable of very good results, especially if open-wire feeders are used to reduce loss. On 28MHz, it consists of two three half-wavelength in-line antennas fed in phase. Here again, results are better with a tuned feeder to minimise losses, although satisfactory results have been claimed for the 10.36m (34ft) stub and 75 ohm feeder.

When using tuned feeders, the feeder taps should be adjusted experimentally to obtain optimum loading on each band using separate plug-in or switched coils. The ATU should be connected to the transmitter with 75 ohm coaxial cable in which a harmonic suppression (low-pass) filter may be inserted.

With tuned feeders and use of lengths other than 32.1m (102ft), operation on the 10, 18 and 24MHz bands is possible, though some relatively high values of SWR can be expected. It should also be noted that the radiation pattern is of the general long-wire type and the position of lobes and nulls will vary with length and frequency.

Instead of simply bringing down a tuned line or ribbon feeder, G5RV arranged that there should be a 10.36m (34ft) matching section of open-wire feeder, which connected to its lower end a length of either 75 ohm impedance twin lead or 80 ohm coaxial cable (see Fig 6.19). If 300 ohm ribbon is used for the matching section, the old type must be cut to a length of 8.5m (28ft) and the slotted variety to a length of 9.3m (30ft 7in). This takes into account the different velocity factors of the two ribbons. Unfortunately, the match to the coaxial or twin feeder at the junction with the lower end of the matching section is only good on 14MHz and 24MHz. If 50 ohm coaxial is used, the VSWR on these bands will rise to 1.8:1.

If a coaxial feeder is being used correctly as a 'flat' un-tuned line, its characteristic impedance will be matched at each of its ends. If this is the case, the coaxial cable may be safely buried along its run with no detriment to its operation. This can be useful when an antenna is located at a considerable distance

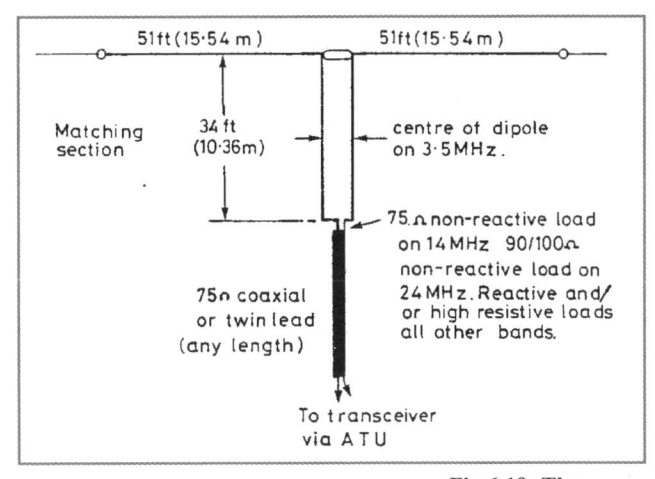

**Fig 6.19:** The G5RV antenna, showing critical dimensions and other details

# HF AMATEUR RADIO

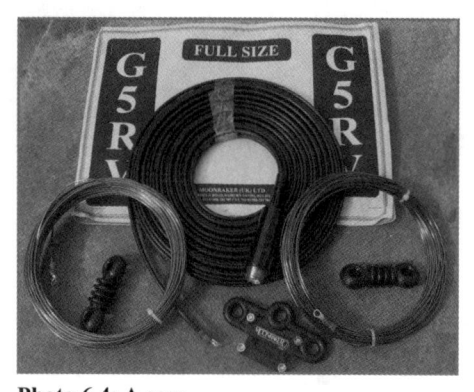

**Photo 6.4: A commercially manufactured G5RV antenna**

from an operating position, but unfortunately the coaxial cable feed of a G5RV antenna must never be buried. The feeder is not correctly terminated and operates with standing waves along its length. It must thus be kept well away from metal and large objects.

## Verticals

Vertical antennas are very popular, as they are able to provide excellent performance without requiring large areas for them to be erected. They also offer an omni-directional radiation pattern, and a low angle of radiation – a distinct advantage when contacting DX stations.

A vertical antenna can be likened to half a dipole in the vertical plane. The other half of the dipole is replaced by the ground as shown in Fig 6.20. In an ideal world, the ground would consist of a perfectly conducting earth but in reality a vertical will operate satisfactorily provided that a good earth connection can be made. Obviously damp soil helps to provide a good earth system.

Unfortunately rocky or sandy environments do not provide a good earth and in these circumstances a vertical antenna may not perform well. An alternative is to have a simulated earth or ground plane. This is normally constructed using a number of quarter-wave radials spreading out from the base of the antenna (see Fig 6.21). Although four radials are normally used, there is no set figure for how many should be used. It is possible to use only one, although performance will be slightly impaired and radiation will be best in the direction of the single radial. Often when a multiband vertical is used, radials of different lengths are used to ensure coverage of all the bands. Some antennas even use loaded radials to shorten their length. The advantage of using a ground-plane system is that it can be mounted above ground level, and this will improve its performance.

The angle at which the radials leave the antenna can be changed. With horizontal radials, the antenna has a low impedance, impedance rises if the radial are angled towards the ground. Often an angle of around 30° gives a good match. In this way, the radials can form part of the guying system, provided that sufficiently strong wire is used. Ultimately, if the radials are angled vertically, then the antenna becomes a vertical dipole and the impedance is approximately 75 ohms, depending on the height above ground.

**Fig 6.20: A vertical antenna**

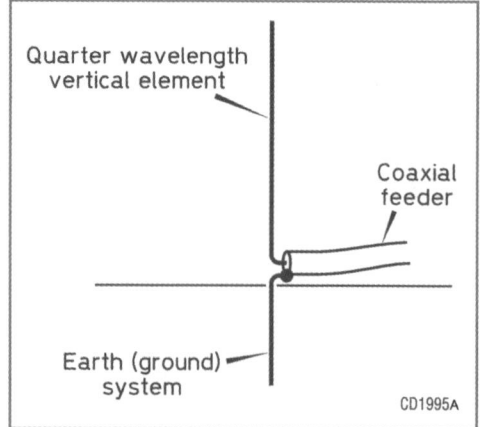

## Trap vertical

It is possible to insert traps into a dipole, and the same is also true for a vertical. In fact, trap verticals are a particularly popular form of antenna. Useable on a number of bands, and not occupying much ground area, they are ideal for people wanting to work DX from a restricted space.

# CHAPTER 6: ANTENNAS

Although it is possible to build a trap vertical, commercially made antennas are normally bought. There is a good selection to choose from and they are robust as they are normally made from tubular aluminium. The traps are generally made as an integral part of the antenna, being slightly wider than the rest of it. The outer sleeve forms a capacitor, while inside there is a coil.

In addition to the normal quarter-wave varieties, a number of half-wave-length antennas are available. These have the advantage that their operation is not dependent on the earth connection, and accordingly they do not need a radial or ground system.

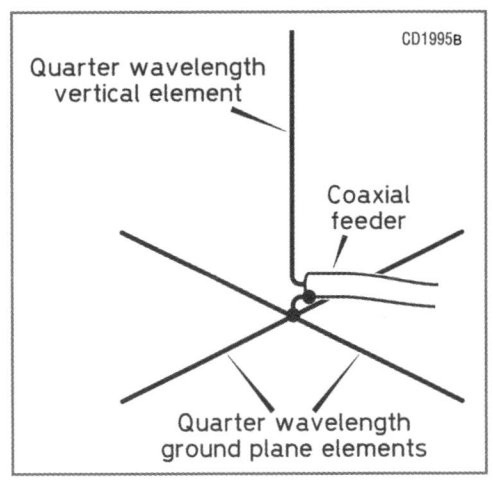

## Beams

While it is not possible for everyone to install a beam antenna, many keen DXers use them because they give improved performance over dipoles, verticals and many other types of antenna.

One of the most popular directive or beam antennas is known as the Yagi. Named after the Japanese researcher who invented it, this type of antenna is used in many areas of radio communications. Most terrestrial television receivers have a Yagi antenna connected to them, and they are also very popular with radio amateurs who need a directional antenna offering gain.

The antenna works because 'parasitic' elements close to a dipole can change its directional characteristics. These extra elements interact with the signal being radiated in such a way that the power is either enhanced or reduced in a particular direction. Elements tend either to 'reflect' or 'direct' the power and, because of this, they are given the names reflector and director.

A Yagi antenna is made up as shown in Fig 6.22. The reflector behind the driven element is made about 5% longer. Only one reflector is used because further ones do not noticeably improve the performance. One or more directors are placed in front of the driven element. The one nearest it is about 5% shorter and any further directors are made slightly shorter than this.

Above, Fig 6.21: A ground plane antenna.
Below, Photo 6.5: An impressive antenna system using a Yagi array

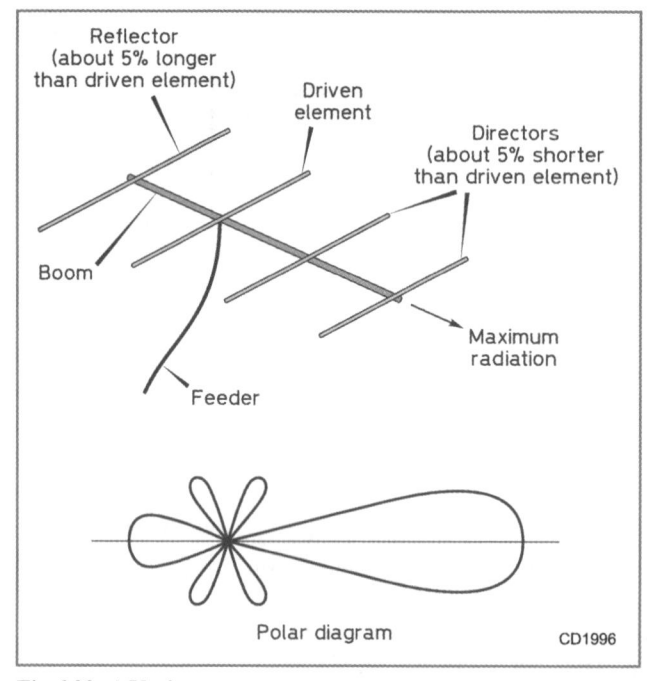

# HF AMATEUR RADIO

**Fig 6.22: A Yagi antenna**

The spacing between the elements will vary from one design to the next. It is usually between a quarter and three-eighths of a wavelength, the exact spacing being chosen to adjust the feed impedance to the required value.

A full-sized Yagi for even the highest-frequency HF bands can be quite large, and as a result it may not be possible to install one. Fortunately, there are some 'minibeams' that are available. While they do not offer the same performance, they may be a good alternative in many instances. Typically, they have a much narrower bandwidth and may not cover a complete band. The level of gain will also be less. This may not be a significant problem, especially if the beam is well situated, free from obstructions.

When using a beam, it should be remembered that the cost of the antenna itself is only part of the total cost. A rotator, possibly long lengths of coaxial cable, and a tower or other mounting will be required. All of these will add to the total cost of the installation.

## Ground or earth connections

A good ground or earth system is essential for the operation of some antennas. An end-fed wire (or in particular a quarter-wave, ground-mounted vertical) requires a good earth as it forms part of the antenna. The nature of the ground itself will play a big part in the way the ground system performs. If it is dry and sandy, it will not be nearly as good as if the area is moist. Whatever the type of soil, a good system must be set down if the earth is to perform well.

**Photo 6.6: A minibeam antenna**

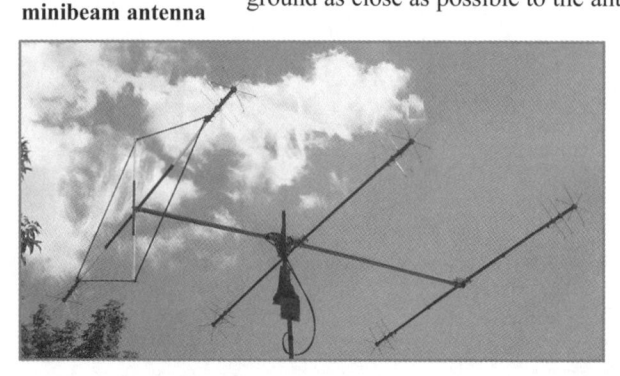

The basic earth connection can be made by driving a long earth rod into the ground as close as possible to the antenna feed point. These rods can be bought from electrical wholesalers relatively cheaply. However, one on its own is unlikely to make a sufficiently good connection, and normally an array of several is required. In addition to earth spikes, spare lengths of copper pipe can be buried (they should not be driven into the ground as the copper is soft and will quickly bend and buckle if hammered in).

Although such an arrangement will provide a reasonable direct current (DC) connection, it may not give a sufficiently good radio frequency performance. To achieve this, a number of radials or counterpoises may be used. These wires are laid out on the ground radiating out from the base of the antenna. Although they can be set out just above the ground, this is seldom practicable and they can be set into it instead. This can be easily done by cutting a thin slit into the ground, and closing up the slit after the wire has been inserted. The slit will soon become invisible, even in a lawn, if it is done carefully.

The radials should generally be about a quarter-wavelength long at the frequency of operation. If the antenna is to cover a variety of bands, then they can be made different lengths to cover all of them. Some stations, particularly short-wave broadcast stations, use more than 100 radials. An amateur station may want to set down an 'earth mat' using multiple radials of different lengths for the bands in use. Another alternative may be to use chicken wire to cover the area around the antenna. Use whatever you can.

**Fig 6.23: An ideal radial earth system for a vertical antenna using many quarter-wave radials for the bands in use**

## Location

It has been said that the ideal location for an antenna would be on an island in the middle of a salt marsh on top of a high plateau. Even then, it would help if the land fell away gently in the direction of the signal. Clearly this is not possible for the majority of radio amateurs! Most of us have relatively small spaces in which to install our antennas. However, do not be discouraged. Many amateurs who regularly contact stations all over the world have very small gardens.

There are naturally a few guidelines to follow. The first is to ensure that, whatever type of antenna is used, it is properly and safely installed. The antenna should also be as far away from the sources of man-made interference as possible. Not only will this reduce the level of received interference from electrical items like motors, fluorescent lights, computers and so forth, but it will also reduce the chances of transmissions interfering with domestic televisions and radios.

It also helps if the antenna is not screened, particularly by buildings, because the wiring and plumbing in them will pick up and absorb the signal. If metalwork is particularly close, then it may detune the antenna.

If power or telephone lines run in the vicinity of the antenna, try to arrange the antenna so that its axis is at right-angles to the lines. In this way, the coupling will be reduced and pick-up of interference will be minimised.

The height of the antenna is also important. In general, it should be as high as possible to enable it to 'see' over the obstructions. Also, antennas that are low only tend to transmit and receive signals that have a high angle of radiation. Low angles of radiation are best for DX use because each reflection from the ionosphere gives the maximum distance – again height gives advantages.

## Safety

Putting metalwork up into the air creates a number of risks and these must be minimised as far as possible. You should take into account safety at every stage of the installation of the antenna, from the initial concept stage right through to its operation. This is particularly important where large, heavy systems are installed.

There are a number of areas where accidents tend to happen. Many accidents are caused by rushing the work. Even if you have a tight deadline, such as the onset of darkness or a contest start, you should still never rush.

Work on HF antennas often involves using a ladder. This should always be securely in place, possibly with someone else at the bottom to make sure the base does not slide. It is also important to ensure that the ladder is at the right angle. When working on uneven ground, a firm base should be created for the ladder using wooden or concrete slabs. Similarly, stepladders should have all four feet firmly on the ground so that the ladder does not rock.

Care should be taken if mains-driven tools are required. They should not be used when it is wet, and an earth-leakage breaker should be used at all times.

When undertaking any work it is essential that someone else is around to give or call for help if the unthinkable happens.

As for the antenna itself, there are a number of rules. First of all, it should not be installed where it could fall onto any power lines. However unlikely this may seem, it has happened before and people have been killed. Similarly, it is probably obvious to state that an antenna should never be erected under power lines.

The proper techniques and fittings should always be used, and the antenna should be inspected periodically. The ravages of the weather will eventually tell on even the most professionally installed system.

If in doubt, it is always best to call in an expert. This is particularly important with towers and large antennas. If even a small tower blows down in the wind, the results could be very serious. If thought is given to safety, there should be no accidents and the antenna can be successfully installed and used, providing many years of effective and safe use.

## Further reading:

*Practical Wire Antennas 2*, Edited by Ian Poole G3YWX, RSGB, 2005.
*Backyard Antennas*, Peter Dodd G3LDO, RSGB, 2000.
*Radio Communication Handbook*, Edited Mike Dennision G3XDV and Chris Lorek G4HCL, RSGB, 2005.

CHAPTER 7

# Bands and Band Plans

*In this chapter:*
- Band plans
- Band summaries
- QRP operation

THERE is an enormous difference in the characteristics of the amateur bands within the short-wave portion of the radio spectrum. From 'Top band' in the MF region to 10m, which is almost a VHF band, there is a tremendous variation in the way signals propagate, the types of antenna that are used, and in some of the circuit techniques that may be employed in the equipment. This makes the short-wave bands particularly interesting. When propagation conditions mean that one band is closed or not able to support long-distance communications, another may be open to places all over the world. So whether you are a night owl or a daytime-only person, there will always be a something of interest on the bands.

One of the secrets of successful operating is knowing where to look and when. Band conditions are always changing, not only over the course of a day, but from day to day as well. Experienced operators know how to tell when conditions are likely to be good. Not only is propagation an issue, but also some of the bands are more popular than others. While this means that there are more stations to contact, it also means there is more interference and competition, so making contacts with stations in rare countries may be more difficult. By knowing which bands to use, the most efficient use can be made of the station and the time available. Band plans also differ between countries – by choosing a position in the band where there is likely to be less competition, the results may be improved. This is where the operator's skill can help make up for not having a high-power station with a large antenna on the top of a hill, and can bring success and a great sense of achievement when using an average station.

## Band plans

The HF amateur bands are subject to band plans in the same way as the VHF bands. In the USA, the planning is mandatory and stations are required under the terms of their licence to keep to them. In the UK and many other countries, band plans are not mandatory but adhering to them makes good sense. Not only will other stations be looking for transmissions of a particular type in a given section of the band, thereby increasing your chances of making a contact, but keeping to the band plan also reduces the levels of interference and makes the best use of the available space. For example, the Morse section of the band is narrower than the section given over to single sideband because a Morse signal takes up much less space than a sideband signal. Operating a sideband signal in the Morse section of the band would cause interference to several Morse stations and accordingly this should not be done. However, it is quite permissible for a Morse signal to be transmitted in the sideband section. Although there would be little point in putting out a Morse CQ call in the sideband section of the band, it is not uncommon for a station to revert to Morse when it is not possible to maintain the contact using single sideband.

Provision is also made for other types of transmission, including data and slow-scan television (SSTV), and there are sections of the bands or spot frequencies reserved for beacons. Care should be taken to avoid transmitting in these reserved areas, particularly the beacon sections as many people will be listening to them and they will not appreciate someone transmitting on the same frequency.

Band plans are not the same around the world for a variety of reasons, including the fact that not all countries have the same band allocations. Summaries of the UK band plans are given below. Up-to-date, comprehensive band plans can be found on the RSGB web site and in the RSGB Yearbook (see 'Further reading' at the end of the chapter).

## Amateur bands

Each of the amateur bands in the MF and HF portion of the radio spectrum has its own very distinct character and is likely to produce signals from different areas and at different times. By using one band instead of another at a particular time, much better results may be achieved.

### 160m (Top Band) (1.81–2.00MHz)

This is a challenging yet interesting band for anyone wanting to make DX contacts and one for the night owls because it only supports long-distance communications at night. However, it can be used for relatively local contacts during the day. At that time, signals are heard via ground wave and, dependent on transmitter powers and antennas, distances of 50 miles or more may be reached. At night, when the D layer disappears, distances increase and it may be possible to hear stations several hundreds of miles away. It is even possible to make transatlantic contacts when conditions are right if sufficiently good antennas are available at both ends.

For very-long-distance contacts the whole of the path must lie in darkness. There can be a significant improvement at dawn and dusk for contacts with the other side of the

**Table 7.1: 160 Metres (Top Band) (1.81 - 2.00 MHz)**

| UK Band Plan Frequencies (MHz) | Usage |
|---|---|
| 1.810 – 1.838 | Morse only |
| 1.838 – 1.840 | Narrow band modes |
| 1.840 – 1.843 | All modes |
| 1.843 – 2.000 | Telephony and Morse |
| 1.843 | QRP |
| 1.960 | DF Contest beacons |

\* narrow band modes should have a bandwidth of no more than 500Hz and include Morse and digimodes globe. These enhancements may only last for 10 to 15 minutes at maximum, and sometimes less.

For shorter paths, like those between Europe and North America, signals peak when it is either sunrise or sunset at one end or the other. Long-distance, north-south paths often peak around midnight. As a general rule, long-distance work improves in winter because of the longer hours of darkness and lower levels of static. As this does not correspond with optimum conditions in the other hemisphere, it means that these signals may be heard at any time of the year.

## 80m (3.5–3.8MHz [3.5–4.0MHz in North America])

Like 160m, this band is shared with other services and can be noisy, especially at night. However, during the day the distances that can be reached are greater than those on 160m. Often stations a few hundred miles away can be heard, making it an ideal band for contacts around the UK. At night, stations from further afield can be heard – distances of over 1000 miles are common, and greater distances can be achieved by those with good antennas. The band comes into its own during the years of the sunspot minimum, but it can perform well at any time.

Propagation along the grey line can produce exceedingly good results, with stations from the other side of the globe being audible at the same strengths as many local stations. However, this may only be short lived and it can be very selective in terms of location.

Most of the SSB DX takes place in a 'DX window' in the top 25kHz of the European band. As a result, this section of the band should be kept clear at all times. This should be observed even when you think there is no possibility of any DX coming through because stations with a good location and good anten-

**Table 7.2: 80 Metres (3.5 - 3.8 MHz [3.5 - 4.0 MHz in North America])**

| UK Band Plan Frequencies (MHz) | Usage |
| --- | --- |
| 3.500 – 3.580 | Morse |
| 3.500 – 3.510 | Priority for inter-continental contacts |
| 3.510 – 3.560 | Preferred section for contest contacts |
| 3.555 kHz | Slow telegraphy centre of activity |
| 3.560 – 3.580 | QRP (3.560 QRP centre of activity) |
| 3.580 – 3.600 | Narrow band modes |
| 3.590 – 3.600 | Narrow band modes – automatically controlled data stations (unattended) |
| 3.600 – 3.800 | Phone (and Morse) |
| 3.600 – 3.620 | All modes – automatically controlled data stations (unattended) |
| 3.600 – 3.650 | Preferred section for phone contest contacts |
| 3.663 | May be used for UK emergency communications traffic |
| 3.690 | QRP SSB centre of activity |
| 3.700 – 3.800 | Preferred section for phone contest contacts |
| 3.735 | Image modes centre of activity |
| 3760 | IARU Region 1 Emergency centre of activity |
| 3.775 – 3.800 | Priority for inter-continental contacts |

nas might just be able to hear DX stations and will not want to suffer from local interference. Stations in North America and other areas of the world have an allocation up to 4.0MHz so it is common to work split frequency, using the DX window below 3.8MHz for European stations and above 3.8MHz for North America etc.

Although antennas for the band can be large, very few stations have beam antennas with high levels of gain. This means that the average station can be quite competitive, especially if the full legal power limit can be achieved.

### 40m (7.0–7.2MHz [7.0–7.3MHz in North America])

The 40m band is a particularly useful band, providing an interesting mix of short-haul DX by day and worldwide communications at night. During the day, stations up to distances of a few hundred miles can often be heard. However, ionospheric absorption limits greater distances. The high angle of skip means that the skip zone is small or non-existent.

Distances increase considerably at night. Stations from further afield are more apparent and, as the skip zone increases, local stations fall in strength. It is a favourite band for many during the low part of the sunspot cycle, being capable of long-haul contacts during the hours of darkness. Again the grey line can produce some spectacular results.

In Europe the band is now 200kHz wide, although the section between 7.100 and 7.200MHz may still have some broadcast stations present. In North America, where frequencies up to 7.3MHz are available, interference from European broadcast stations (to whom this portion is allocated in Europe) can be a problem.

This band can be a good hunting ground for those with medium powers and average antennas. Comparatively few people use directive antennas and this means that those with average stations are at less of a disadvantage. Trap verticals, provided they are operated against a good earth or ground-plane system, can give a good account of themselves, allowing stations all over the world to be contacted.

Table 7.3: 40 Metres (7.0 - 7.2 MHz [7.0 - 7.3 MHz in North America])

| UK Band Plan Frequencies (MHz) | Usage |
|---|---|
| 7.000 – 7.035 | Morse only |
| 7.030 | QRP centre of activity |
| 7.035 – 7.040 | Narrow band modes |
| 7.038 – 7.040 | Narrow band modes – automatically controlled data stations (unattended) |
| 7.040 – 7.043 | All modes – automatically controlled data stations (unattended) |
| 7.043 | Image modes centre of activity |
| 7.043 – 7.200 | All modes excluding digimodes |
| 7.045 | May be used for UK emergency traffic. |
| 7.060 | IARU Region 1 centre of activity for emergency traffic |

## 30m (10.100–10.150MHz)

This band was released for amateur use after the World Administrative Radio Conference held in 1979 (WARC 79). It is still not very widely used but is capable of giving good results. It is very similar in character to the 40m band, being only slightly higher in frequency.

The band is capable of giving DX contacts for most of the day, although it is generally better at night. For most of the time, a skip zone is apparent, except at the peak of the sunspot cycle. The level of absorption is less than on 40m, and during the night distances increase. Again conditions are enhanced by grey line and dusk or dawn conditions. During periods of the sunspot minimum, when ionisation levels are lower, absorption is sufficiently low to allow long-distance contacts throughout the day.

Like the 40m band, this and the other WARC bands are good bands for the DXer who does not have a really big station. Few of the common directional Yagi antennas work on this band and some stations may still be using linear amplifiers that cannot operate here. As a result, those with average stations are at less of a disadvantage.

Due to the small size of the band and the high level of commercial activity (because it is shared with other services), most of the operation is in Morse. The IARU for Region 1 have recommended that contests and phone operation should be excluded from the band.

**Table 7.4: 30 Metres (10.100 - 10.150 MHz)**

| UK Band Plan Frequencies (MHz) | Usage |
| --- | --- |
| 10.100 – 10.140 | Morse |
| 10.116 | QRP centre of activity |
| 10.140 – 10.150 | Narrow band modes |

NB: Automatically controlled data stations (unattended) should avoid the use of the 10 MHz band

## 20m (14.0–14.35MHz)

This is the main long-haul band for radio amateurs, reliably giving the possibility of long-distance contacts during all phases of the sunspot cycle. During the day, stations up to about 2000 or 3000 miles away can be heard when conditions are good, and there are virtually always stations between 500 and 1500 miles which can be heard. The band normally closes at night during the winter and during the sunspot minimum, but during the summer and the sunspot maximum it will remain open most of the night. Spring and autumn normally produce good results, with stations from the other hemisphere being heard with ease at various times of the day.

Over the course of a day, signals can be heard from all over the world. In the early morning signals arrive from the east. When these signals fade out, more local signals will become prominent, and there may be openings to the west as the Sun rises in that direction. As the afternoon wears on, openings further west may arise. There may also be openings to the other side of the globe again as their morning approaches. In the evening, as the levels of ionisation fall, the local signals will fall in strength, leaving long-distance stations to the west.

HF AMATEUR RADIO

**Photo 7.1: An efficient HF antenna system that could be used on 20m**

From the UK, stations in Canada and on the north-east coast of the USA are heard first in the afternoon or evening, and then the skip can be seen to move southwards, encompassing the southern states and then down into the Caribbean and South America.

Being the mainstay DX band, 20m is often crowded and, when any rare stations appear, competition is often great, making high powers and good antennas much more imperative. Some of the big stations run powers of the order of a kilowatt (where licensing conditions permit) and use three-element Yagi antennas at a height of around 60ft (20m). Nevertheless, it is still possible to make many good contacts. Good operating techniques and listening for stations before the pile-ups become too large enable contacts to be made with DX stations. When the conditions are good, it will be necessary to decide whether to stay with a pile-up or move on to find if there are any other DX stations with whom contact is more likely.

## CHAPTER 7: BANDS AND BAND PLANS

**Table 7.5: 20 Metres (14.0 - 14.35 MHz)**

| UK Band Plan Frequencies (MHz) | Usage |
|---|---|
| 14.000 – 14.070 | Morse |
| 14.000 – 14.060 | Contest preferred |
| 14.055 | QRS (slow telegraphy) centre of activity |
| 14.060 | QRP centre of activity |
| 14.070 – 14.099 | Narrow band modes |
| 14.089 – 14.099 | Narrow band modes – automatically controlled data stations (unattended) |
| 14.099 – 14.101 | Beacons only |
| 14.101 – 14.112 | All modes including automatically controlled data stations (unattended) |
| 14.112 – 14.350 | All modes |
| 14.112 – 14.125 | All modes except digimodes |
| 14.125 – 14.300 | Preferred section for SSB contests |
| 14.195 +/– 5kHz | Priority for DXpeditions |
| 14.230 | Image centre of activity |
| 14.285 | QRP centre of activity |
| 14.300 | Global emergency centre of activity |

## 17m (18.068–18.168MHz)

Like the 30m band, this one was released for amateur use after WARC 79 and some older transceivers may not cover it. It is very much a half-way house between 15 and 20m. Although rather narrow, it is still very popular and well worth investigating when conditions look promising.

The band can offer some excellent opportunities for contacting DX stations. Although beam antennas are available for it, most stations still use dipoles as those with beams may use them for the more traditional DX bands of 10, 15 and 20m, thereby limiting the number of strong stations. However, more antennas are appearing for the WARC bands with the result that more people are using these frequencies.

**Table 7.6: 17 Metres (18.068 - 18.168 MHz)**

| UK Band Plan Frequencies (MHz) | Usage |
|---|---|
| 18.068 – 18.095 | Morse |
| 18.095 – 18.109 | Narrow band modes |
| 18.105 – 18.109 | Narrow band modes – automatically controlled data stations (unattended) |
| 18.109 – 18.111 | Beacons only |
| 18.111 – 18.168 | All modes |
| 18.111 – 18.120 | All modes – automatically controlled data stations (unattended) |
| 18.160 | Global emergency centre of activity |

# HF AMATEUR RADIO

### Table 7.7: 15 Metres (21.0 - 21.45 MHz)

| UK Band Plan Frequencies (MHz) | Usage |
|---|---|
| 21.000 – 21.070 | Morse |
| 21.055 | QRS centre of activity |
| 21.060 | QRP centre of activity |
| 21.070 – 21.110 | Narrow band modes |
| 21.090 – 21.110 | Narrow band modes – automatically controlled data stations (unattended) |
| 21.110 – 21.120 | All modes (excluding SSB) |
| 21.120 – 21.149 | Narrow band modes |
| 21.149 – 21.151 | Beacons only |
| 21.151 – 21.450 | Phone (and Morse) |
| 21.285 | QRP centre of activity |
| 21.340 | Image centre of activity |
| 21.360 | Global emergency centre of activity |

### 15m (21.0–21.45MHz)

This band is more variable than 20m, being affected more by the sunspot cycle. During the peak, it is open during the day and well into the night when it will support propagation over many thousands of miles. When the band is open like this, strengths are usually better on it than 20m because of the effect of the D layer is less. Conditions are usually not quite so good in the early morning, improving as the day progresses. During the sunspot minimum, few stations may be heard during the day and none at night.

At the top of the band is the 13m broadcast band. Tuning up into this will give a quick indication of whether the amateur band may be open.

### 12m (24.890–24.990MHz)

This band is greatly affected by the sunspot cycle and has many similarities with 10m. Although it may just support propagation when the latter cannot, it will follow very much the same pattern.

Like 17m, this band also is quite narrow but worth investigating. Also, there are few stations using beam antennas on this band and this makes it a good hunting ground.

### Table 7.8: 12 Metres (24.890 - 24.990 MHz)

| UK Band Plan Frequencies (MHz) | Usage |
|---|---|
| 24.890 – 24.915 | Morse |
| 24.915 – 24.929 | Narrow band modes |
| 24.925 – 24.929 | Narrow band modes – automatically controlled data stations (unattended) |
| 24.929 – 24.931 | Beacons only |
| 24.931 – 24.990 | All modes |
| 24.931 – 24.940 | All modes – automatically controlled data stations (unattended) |

## 10m (28.0–29.7MHz)

This is the highest-frequency band in the short-wave (HF) portion of the spectrum. During the sunspot minimum, it may only support ionospheric propagation via Sporadic E which occurs mainly in the summer months. This gives propagation over distances of 1000 miles or so.

At the peak of the sunspot cycle, it gives excellent possibilities for long-distance contacts, producing very strong signals. This band is well known for enabling stations with low powers and poor antennas to make contacts over great distances as ionospheric absorption is less than on the lower-frequency bands. In general, propagation on these frequencies requires that the path is in daylight. Despite this, at the peak of the sunspot cycle, the band may remain open into the night, although it will eventually close.

There are a large number of beacons active on frequencies between 28.175 and 28.30MHz. Many of them are outside the allocated beacon band and care should be taken not to operate on top of them.

Although much of Morse operation is centred around the bottom of its allocation, check the upper part of this allocation, especially during contests, as levels of

Table 7.9: 10 Metres (28.0 - 29.7 MHz)

| UK Band Plan Frequencies (MHz) | Usage |
| --- | --- |
| 28.000 – 28.070 | Morse |
| 28.055 | QRS centre of activity |
| 28.060 | QRP centre of activity |
| 28.070 – 28.190 | Narrow band modes |
| 28.120 – 28.150 | Narrow band modes – automatically controlled data stations (unattended) |
| 28.190 – 28.225 | Beacons |
| 28.190 – 28.199 | Regional time shared beacons |
| 28.199 – 28.201 | World wide time shared beacons |
| 28.201 – 28.225 | Continuous duty beacons |
| 28.225 – 29.300 | All modes |
| 28.225 – 28.300 | Beacons |
| 28.300 – 28.320 | All modes – automatically controlled data stations (unattended) |
| 28.360 | QRP centre of activity |
| 28.680 | Image centre of activity |
| 29.200 – 29.300 | All modes – automatically controlled data stations (unattended) |
| 29.210 | UK Internet voice gateway (unattended) |
| 29.290 | UK Internet voice gateway (unattended) |
| 29.300 – 29.510 | Satellite downlinks |
| 29.510 – 29.520 | Guard channel |
| 29.520 – 29.700 | All modes |
| 29.520 – 29.550 | FM simplex (10 kHz channels) |
| 29.530 | UK Internet voice gateway (unattended) |
| 29.560 – 29.590 | FM repeater inputs (RH1 – RH4) |
| 29.600 | FM calling channel |
| 29.610 – 29.650 | FM simplex (10 kHz channels) |
| 29.630 | UK Internet voice gateway (unattended) |
| 29.660 – 29.700 | FM repeater outputs (RH1 – RH4) |

activity force stations further up the band. Under these conditions, competition at the top may be less than further down the band, giving a greater possibility of making contacts.

Activity in the SSB portion of the band is often concentrated between the beacon section and 28.60MHz and a little above. However, it is again worth taking a look above this, particularly in contests.

Stations using low-power FM may be heard towards the top of the band. The recommendation is that FM activity should take place between 29.60 and 29.69MHz, with 29.60MHz as the calling frequency. There are some repeaters in the USA with outputs at 29.62, 29.64, 29.66 and 29.68MHz with inputs 100kHz lower.

## QRP operation

There is a growing band of people who enjoy building and operating low-power equipment. Most of the operation is on Morse, for two reasons. The first is that the equipment for this mode is often simpler, making home construction much easier. The second is that it is possible to copy Morse at lower signal strengths than single sideband or any other form of speech transmission. This makes long-distance contacts easier to make using this mode.

QRP frequencies are reserved in each of the HF bands for QRP operation. These frequencies, which are shown in the band plans, should be avoided by high-power stations to allow those using low power to have the minimum amount of interference and hence have the best chance of making their contacts.

The definition of what QRP actually is varies somewhat. In the USA, it is often considered to constitute stations operating under 100W. The UK-based G-QRP Club defines QRP as being power levels under 5W DC input. IARU Region 1 defines QRP as under 10W input and QRPp as under 1W input.

Making contacts using low powers can be particularly rewarding. However, low power may not be as much of a disadvantage as might be first thought. Reducing the power from 400W down to 4W output represents a reduction of 20dB. A figure of 6dB is generally taken to be equivalent to an 'S' point and therefore this power reduction represents a reduction of just over three 'S' points. In other words, if a station running the full UK legal output of 400W (26dBW) was being received at S9 and it were then to reduce the power to just 4W (6dBW), it would still be copied at around S6. While a QRP station might not be able to operate through many pile-ups, the figures show that it should still be possible to make plenty of contacts.

Along with using low power, QRP operators usually take pride in building much of their equipment themselves. This also brings a great sense of achievement. Some people enjoy making their equipment by buying the components and making the equipment from scratch, while others use some of the growing number of kits that are avail-able. Some of these are relatively simple but there are a number of full single sideband transceivers available.

## Further reading:

*Amateur Radio Operating Manual,* 6th edn, ed Don Field, G3XTT
    RSGB, 2004.
*RSGB Yearbook* (published annually by the RSGB).

CHAPTER 8

# On the Bands

*In this chapter:*

- Starting out
- Basic contacts
- DX techniques
- Getting through in a pile-up
- QSL cards
- Awards
- Contests

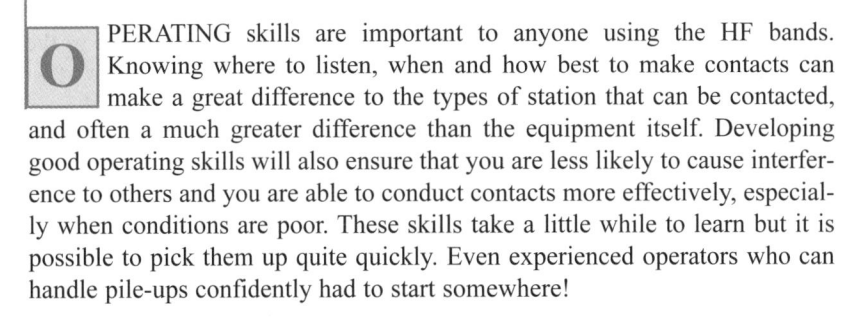

OPERATING skills are important to anyone using the HF bands. Knowing where to listen, when and how best to make contacts can make a great difference to the types of station that can be contacted, and often a much greater difference than the equipment itself. Developing good operating skills will also ensure that you are less likely to cause interference to others and you are able to conduct contacts more effectively, especially when conditions are poor. These skills take a little while to learn but it is possible to pick them up quite quickly. Even experienced operators who can handle pile-ups confidently had to start somewhere!

## Starting out

To find out about operation on the HF bands, it is worth taking some time to listen to them. Indeed, one of the best apprenticeships for anyone wanting to operate on the HF bands is to spend some time as a short-wave listener. In this way, it is possible to find out how typical contacts are made, what is said and how the information is put over in a concise way so that the other station is totally aware of what is happening, even when there are high levels of interference – it is essential to know how to handle a contact under difficult conditions. However, to start out it is useful to know the basic format for a contact.

## Basic contacts

Many of the contacts that take place on the short-wave bands are what are termed 'rubber stamp' contacts. These are a good starting place for many people, including those who do not speak English as their first language, as it is relatively easy to make a contact using a minimal vocabulary. However, many people like to talk about far more than the topics covered by the basic contact. Often, they will discuss technical matters or describe the part of the world where they live. Many will want to have long conversations once they are familiar with operating but first of all they need to know the basic elements of a contact.

On the HF bands, a contact will often commence with a CQ call (a general call to all stations). A formula known as 'three times three' is a good starting point. Here the letters 'CQ 'are repeated three times, and then the callsign, usually spoken using phonetics, is repeated three times. This whole procedure is repeated three times. In this way, the call is kept to a reasonable length and anyone listening is able to gain the callsign and the fact that a contact is wanted.

Any station listening who wants a contact can then respond when invited to do so. He will normally give his callsign a couple of times using phonetics and then invite the other station to transmit.

If the first station hears the caller and responds, he will announce the callsigns and then normally wish him 'good day' and give a signal report. This tells the other station how he is being received and gives information about his station's performance, so that if conditions are difficult, the con-tact can be suitably tailored. After he gives the report, it is normal for him to give his name and location. Then the callsigns will be given and transmission handed over.

The second station will follow a similar format, giving a report, his name and location. On the next transmission, information about equipment in the station – the transmitter, receiver or transceiver and the antenna – is often given. Details of the weather are also often mentioned.

On the third transmission, details about exchanging QSL cards may be given and then the two stations may sign off. Again, callsigns will be given at the beginning and end of each transmission.

Once a contact has finished, it is perfectly permissible to call one of the stations. Normally, the frequency 'belongs' to the person who called CQ initially, but if the other station is called he may ask to keep the frequency or move off to another one.

Giving callsigns at the beginning and end of each transmission may seem somewhat formal, but it fulfills legal requirements to identify the station and also lets the other station know exactly what is happening and when he is expected to transmit. Using good operating technique is very important and helps contact to be maintained with the minimum confusion, especially when conditions are poor or interference levels are high. One of the keys to becoming a good operator is to let the other station know exactly what you are doing and not leave him guessing.

When a station in a very rare location is on the band, contacts are usually kept very much shorter to enable as many contacts as possible to be made, and this also applies to contest operating. Usually the contact will consist of just the callsigns of the stations and then a report. Speedy operating is of the essence under these conditions to ensure that others are not kept waiting.

Contacts are very similar during contests, except that the report often has a serial number after it, e.g. 591234 where 1234 is the number required in the contest. This may be the serial number of the contact, or another number, as required by that particular contest.

### Table 8.1: A typical rubber stamp HF contact

CQ CQ CQ FROM GOLF THREE YANKEE WHISKEY XRAY, GOLF THREE YANKEE WHISKEY XRAY, GOLF THREE YANKEE WHISKEY XRAY,
CQ CQ CQ FROM GOLF THREE YANKEE WHISKEY XRAY, GOLF THREE YANKEE WHISKEY XRAY, GOLF THREE YANKEE WHISKEY XRAY,
CQ CQ CQ FROM GOLF THREE YANKEE WHISKEY XRAY, GOLF THREE YANKEE WHISKEY XRAY, GOLF THREE YANKEE WHISKEY XRAY,
AND G3YWX IS STANDING BY FOR A CALL.

G3YWX, G3YWX, THIS IS VP8ANT, VICTOR PAPA EIGHT ALPHA NOVEMBER TANGO, VICTOR PAPA EIGHT ALPHA NOVEMBER TANGO

VICTOR PAPA 8 ALPHA NOVEMBER TANGO THIS IS GOLF THREE YANKEE WHISKEY XRAY. GOOD MORNING OLD MAN AND THANK YOU FOR THE CALL. YOUR REPORT IS FIVE AND NINE, FIVE AND NINE. MY NAME IS IAN, INDIA ALPHA NOVEMBER AND THE QTH IS LONDON, LIMA OSCAR NOVEMBER DELTA OSCAR NOVEMBER. HOW DO YOU COPY? VP8ANT, G3YWX LISTENING

GOLF THREE YANKEE WHISKEY XRAY, THIS IS VICTOR PAPA EIGHT ALPHA NOVEMBER TANGO. THANK YOU FOR THE REPORT AND COMING BACK TO MY CALL IAN. YOUR REPORT IS ALSO FIVE BY NINE, FIVE BY NINE AND THE NAME HERE IS RICHARD, ROMEO INDIA CHARLIE HOTEL ALPHA ROMEO DELTA AND THE LOCATION IS IN ANTARCTICA, ALPHA NOVEMBER TANGO ALPHA ROMEO CHARLIE TANGO INDIA CHARLIE ALPHA. BACK TO YOU TO SEE HOW YOU COPY, VP8ANT STANDING BY FOR G3YWX

VP8ANT, G3YWX RETURNING. THANK YOU FOR THE REPORT RICHARD. THE STATION HERE IS A HOMEBUILT AND RUNNING ABOUT 100 WATTS AND THE ANTENNA IS A DIPOLE. THE WEATHER IS FINE HERE, ABOUT 20 DEGREES. BACK TO YOU. VP8ANT, G3YWX LISTENING

G3YWX THIS IS VP8ANT. THANK YOU FOR THE INFORMATION ABOUT YOUR STATION. HERE I AM USING A COMMERCIAL STATION RUNNING ABOUT 200 WATTS AND THE ANTENNA IS A MULTIBAND DIPOLE. THE WEATHER HERE IS COLD, ABOUT MINUS 30 DEGREES BUT SUNNY. SO BACK TO YOU. G3YWX, VP8ANT LISTENING FOR YOU.

VP8ANT, G3YWX. THANK YOU FOR THE INFORMATION RICHARD. I WOULD LIKE TO EXCHANGE QSL CARDS WHICH I WILL DO VIA THE BUREAU. I AM NOW QRU SO I WILL WISH YOU BEST 73S AND LOOK FORWARD TO ANOTHER CONTACT. VP8ANT, G3YWX LISTENING

G3YWX, VP8ANT RETURNING. ALL FINE IAN, AND I AM HAPPY TO EXCHANGE QSL CARDS. 73S AND THANK YOU FOR THE CONTACT. G3YWX, VP8ANT SIGNING AND WISHING YOU A GOOD DAY.

73S RICHARD AND THANK YOU FOR THE CONTACT, G3YWX CLEAR.

# HF AMATEUR RADIO

### Table 8.2: A typical pileup contact

CQ, CQ, CQ, VICTOR PAPA EIGHT ALPHA NOVEMBER TANGO

GOLF THREE YANKEE WHISKY X-RAY

GOLF THREE YANKEE WHISKY X-RAY, 59 (There is no need for VP8ANT to repeat his callsign, all he needs to do is identify the station he is calling and give a report)

GOLF THREE YANKEE WHISKY X-RAY, 59 (G3YWX identifies himself against the interference and give a report)

GOLF THREE YANKEE WHISKY X-RAY, THANK YOU, VICTOR PAPA EIGHT ALPHA NOVEMBER TANGO QRZ (and he listens for the next caller)

### Table 8.3: A typical Morse contact

CQ CQ CQ, DE G3YWX G3YWX G3YWX CQ CQ CQ, DE G3YWX G3YWX G3YWX CQ CQ CQ, DE G3YWX G3YWX G3YWX AR K

G3YWX DE G3XDV G3XDV AR KN

G3XDV DE G3YWX GM OM ES TNX FER CALL UR RST 599 599 = NAME ERE IS IAN IAN ES QTH STAINES STAINES = SO HW CPI? AR G3XDV DE G3WX KN

G3YWX DE G3XDV FB OM ES TNX FER RPRT UR RST 599 599 = NAME IS MIKE MIKE ES QTH NR LONDON LONDON = SO HW? AR G3YWX DE G3XDV KN

G3XDV DE G3YWX FB MIKE ES TNX FER RPRT = TX ERE RNG 30 WATTS ES ANT VERT = WX FB SUNNY ES ABT 23 C = SO HW CPI? AR G3XDV DE G3YWX

G3YWX DE G3XDV R R AGN IAN = RIG ERE RNG 100 WATTS ES ANT DIPOLE UP 10 METRES = WX WET ES COLD ABT 15 C = G3YWX DE G3XDV KN

G3XDV DE G3YWX FB MIKE ES UR RIG DOING FB. ERE QRU = QSL VIA BURO = 73 ES HPE CUAGN SN AR G3XDV DE G3YWX KN

G3YWX DE G3XDV R R QRU ALSO = QSL FB VIA BURO = SO TNX FER QSO 73 ES BCNU AR G3YWX DE G3XDV VA

G3XDV DE G3YWX FM 73 ES BCNU AR G3XDV DE G3YWX VA

Note = sign used as a full stop or break

---

Similar contacts can be made when using Morse, the main difference being that far more abbreviations are used to ensure that the speed at which information can be passed is as quick as possible.

When there are contests, the contacts are made as short as possible in the same way they are for "phone" contacts. This saves time and makes it possible to make more contacts. Table 8.4 shows a typical contact when QSOing with a rare DX station.

# CHAPTER 8: ON THE BANDS

The main difference between this and a contest contact is that the contest report will consist of the report and serial number. People may also call CQ TEST; a shortened form of CQ contest.

## Tricks of the trade

When some leading DX operators were asked what the key to their success was, virtually all of them said it was listening. While there is a great temptation to put out a CQ call and expect the rare and exotic DX station to reply, this seldom happens. Instead, listen around the bands to find out what is going on, so you can find out if there is anything of interest on the band and see what conditions are like. If you spend all the time transmitting, then you can't be searching out the interesting DX.

Even when you spend plenty of time listening there are a number of easy ways to help find the more interesting stations faster. Listen to find out if the operator is speaking with a different accent to the rest of the stations on the band, or if he sounds different in some way. If so, he may be from a different part of the world. Also, when tuning up and down the band, it is worth listening out for a 'pile-up' where many stations are trying to call a particular station. The station under the pile-up is almost certainly going to be of interest.

Also listen out for signals that sound different in other ways. For example, those that come across the Polar Regions often have a 'watery' or 'fluttery' sound to them, indicating they are some distance away. It is also possible to pick up some interesting stations as the band is closing at night. At this time, interference levels may be less from short-skip stations and competition from stations to the east will also be less. Furthermore, long-distance stations are often heard and there is less competition contacting them.

**Table 8.4: A typical rare DX station contact**

| |
|---|
| CQ CQ CQ DE VP8ANT |
| DE G3YWX K |
| G3YWX 599K |
| DE G3YWX 599 K |
| R 73 DE VP8ANT QRZ |

## Information is essential

While the greatest weapon in the DXer's armoury is listening, information is also vitally important. This can be general knowledge, possibly about technical issues, or it may be up-to-the-minute information about what is going on.

A good knowledge of how signals propagate on each band and the way in which they vary over the day is vital. Predictions about the conditions on the bands can be ascertained from magazines like RadCom as well as from the the Internet (see Chapter 2 – 'Propagation'). Details of the A and K indices can be picked up and judgements made about the likely conditions. While propagation predictions are much better these days, it is always valuable to listen around the bands to see if they are as forecast. Don't just check the one that is most likely to be the best, but others as well.

It also helps to know which stations are likely to be on the bands. Information or 'intelligence' about the activity of DXpeditions and other DX stations, such as their operating times and frequencies, is all useful. For non-DXpedition stations, the routine of daily life often means that they tend to operate the same time each day and they may also have their favourite frequencies. It has been said that DXers have an insatiable appetite for information (and even rumour) about DX stations. There are a number of ways in which this information can be gained, and one of the most obvious is from magazines like RadCom. Every month there is an HF column giving details of the DX operations that are on the bands and information about forthcoming operations. This is very useful for getting advance warning about interesting operation.

The Internet is also a very good source of information. The RSGB's own web site (www.rsgb.org.uk) not only provides DX news but also links to some excellent web sites. E-ham.net (www.eham.net) is another very useful site. It is worth adding these to the favourites list on your browser. Another useful site with links is www.radio-electronics.com.

One particularly good method of quickly finding out what is happening on the bands is by using packet radio. There is a system known as the DX Packet Cluster that enables people to see what DX is taking place or has been on the bands and on what frequencies. Its advantage is that it is almost real-time. Some years ago, Dick Newell, AK1A, developed a form of real-time electronic conferencing using packet radio which allowed all the stations that were connected to see everyone else's messages. The idea was taken up by the DX community who saw the advantages the system could offer for passing information around quickly about any DX stations that had appeared on the bands or other relevant news. Now the DX Packet Cluster has become one of the most useful sources of real-time information available to the DXer. There are also some DX Cluster web sites.

The network of stations in the cluster is growing and is present in most countries that have a significant amateur population. For example, most of England is covered and there are links to similar clusters on mainland Europe. There are also clusters in the USA.

Once information is published on a cluster it rapidly spreads throughout the system and can be seen by every station connected to it. The system also caters for short one-line announcement messages, 'talk through' to individual stations, conventional store-and-forward mail and bulletins, WWV propagation data, as well as access to several on-line databases. The system stores data so it is possible to log in and find out what has been happening over the past few hours. Information can be sorted and this can help in looking for patterns of operation that might help in predicting the best time to look for a particular station or country.

All that is needed to use the DX Cluster is an ordinary packet station. You can then connect with the nearest cluster node, direct or via a conventional packet node. Packet stations can be set up quite easily these days, possibly using equipment that is already available in the shack – this makes it a very attractive and efficient way of finding out what is happening on the bands.

Often DXers will alert one another to rare DX by telephone. In fact, some have managed to enlist the help of local short-wave listeners so that if they hear anything of interest they quickly give them a call.

Another useful source of information are DX nets. These change from time to time, so their details are not included here, but details do appear periodically in magazines' HF columns. You might want to simply listen to all the latest information, or you may want to join in and share your experiences and information.

## Pile-ups

When listening around the bands, a pile-up is an almost sure-fire indicator that there is an interesting station around. Pile-ups require a little extra operating skill if you are to get through without causing interference to others. Again, the best advice is to listen. Find out the callsign of the station, listen to hear what the station sounds like, so that he can be identified easily from all the other stations calling. Find out his mode of operation. By doing this, you will know the best time to call.

Many very rare stations operate on split frequencies. The reason for this is that when the pile-ups get very large, the DX station needs to have to access to a frequency where he won't be swamped by interference from those trying to call him. Typically SSB stations may listen 10 and 20kHz higher in frequency or they may listen over a band of frequencies. Those using Morse are often about 2kHz from their transmit frequency.

The other advantage of this approach is that it can spread the stations out, making it easier to pick out a single station from the enormous cacophony of the pile-up. When this mode of operation is employed, those stations who have two VFOs on their transceiver can listen to the stations being contacted and follow the tuning pattern for the DX station, and thereby they know the best frequencies to use.

When calling a station in a pile-up, it is also necessary to get one's call in at exactly the right time after the previous contact has finished and when the DX station is listening. Quick and exact operating is the name of the game. Give your callsign quickly and clearly. However, in the excitement, it is necessary to be careful not to call too soon or out of place, thereby causing interference.

Some pile-ups may be so big that it is just not worth calling, as you may spend many hours calling to no avail. Instead, it may be worth making a note

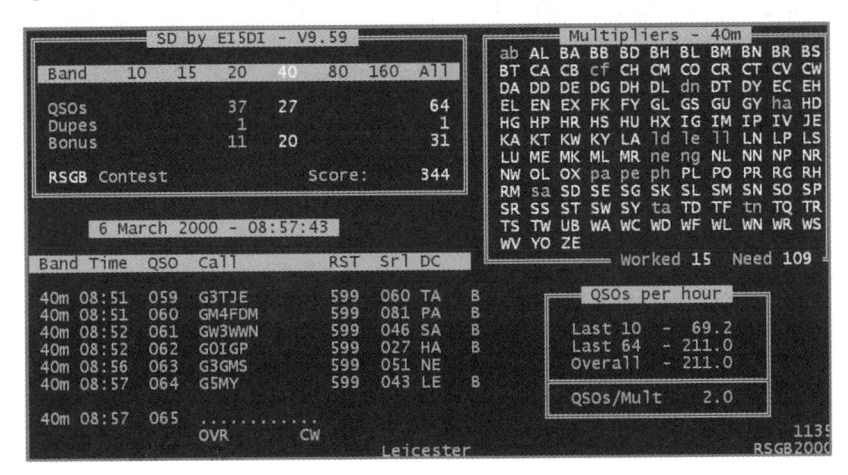

Photo 8.1: Screen shot of a typical contest contact logging software package

Photo 8.2: A typical QSL card

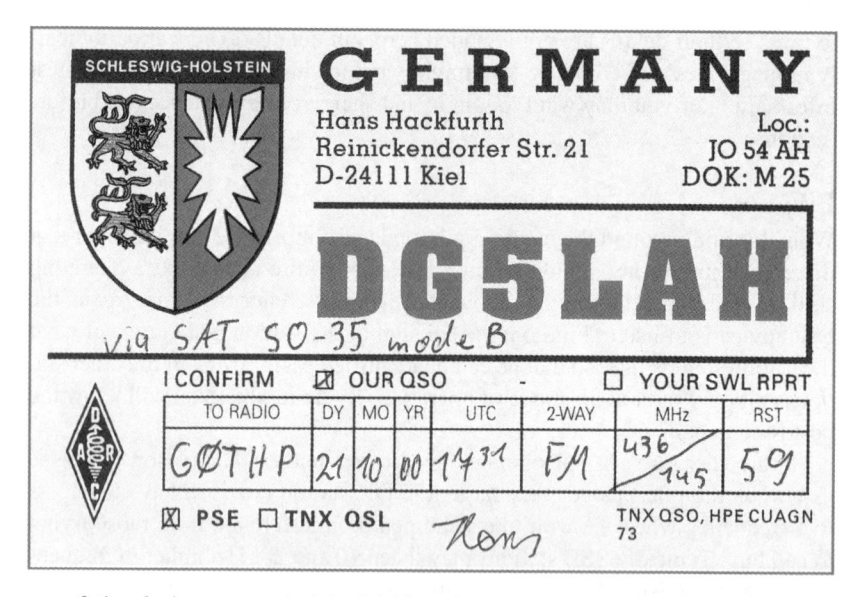

of the frequency and then looking for other DX stations on the band, returning later when propagation conditions may be more favourable. Remember, though, it is often worth making a few calls just in case it is possible to get through. Some of the most satisfying contacts come when you did not expect to get through or when you have been rewarded after spending a long while calling.

## Automatic logging programs

Many of the leading DX operators use computer-based logging software packages. While some people still like to use more traditional paper-based logbooks, there are many advantages to using a computer-based system. Not only is it possible to make duplicate copies of a log, for example for sending off after a contest, but it is possible to check on countries contacted, confirmed and so forth. These packages can provide a great number of statistics very easily. On top of this, many packages can provide partial automatic operation on Morse, sending messages like 'CQ' and the standard replies, making it much easier to operate for long periods of time. They can even be used for voice operation via PC sound cards, if only for calling CQ. There are a number of packages available – some are more suited to home operation, whereas others are aimed at contest operators. It is worth talking to others for advice on the best package to buy.

## QSL card collecting

Many people enjoy confirming a contact with a QSL card. While few send cards for every contact these days, the volume passing through the QSL bureaux is still very high. The RSGB alone processes around 1.5 million cards a year. Once received, they can be displayed on the shack wall to show what DX has been worked, or collected to prove contacts have been made when applying for operating awards.

# CHAPTER 8: ON THE BANDS

**Photo 8.3: A selection of QSL cards from around the world**

QSL cards comes in many forms. Some are printed in a single colour, although many are multi-coloured or have photos on them. Obviously, these cost considerably more to print, but with modern printing processes, they are not as costly as they used to be.

Although there is no hard and fast rule about how a card should look or exactly how the information should be presented, it should contain certain details about the contact. The card should obviously have the callsign of the station printed prominently on it. The address or location should also be

# HF AMATEUR RADIO

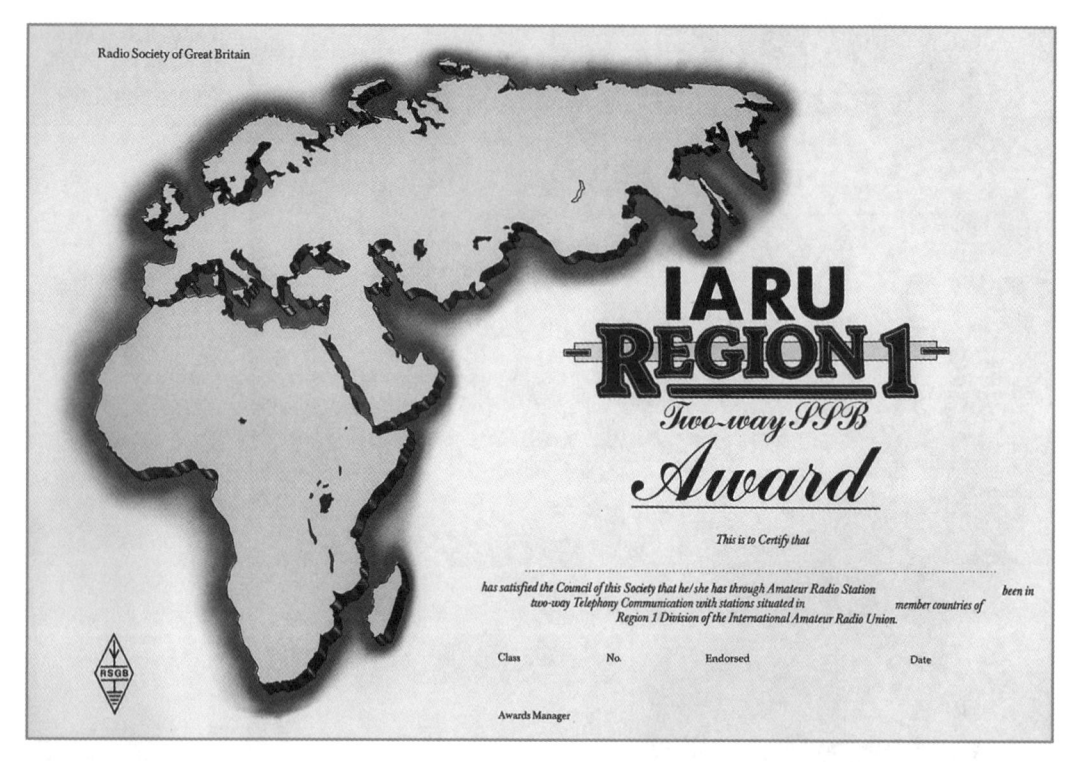

**Photo 8.4: A typical operating award (use one from HFAR1)**

included, along with the name of the owner of the station. Space is then provided to fill in details of the particular contact, including the callsign of the station with whom the contact was made, the date and time (usually in GMT or z). The frequency band or frequency are the mode are also required. The report that was given in the contact should be included. Details of the equipment are very useful. Finally there is normally space to give details of whether a return QSL card is wanted, and the route that can be used. Typically, the wording is "PSE / TNX QSL DIRECT / VIA BURO". Finally there is a space for the operator's signature.

Most people will send their QSL cards via the bureau. This is by far the cheapest option but return cards can take months or years before they are received. Cards can be sent directly for those chasing awards or wanting to have a rare station or country confirmed. When sending cards directly, it is normal to send the return postage. For foreign destinations, dollar bills can be used, although the normal way is to use International Reply Coupons which can be exchanged for return-rate, surface-mail postage. These can be bought from main post offices in the UK but not from sub post offices. Unfortunately, they are rather expensive and as a result many people do not cash them for return postage, instead either selling them on at a cheaper rate, or re-using them for a card they want. This is perfectly acceptable as they do not have an expiry date and as a result have almost become an international form of currency.

One of the major problems in sending cards directly is finding the address of the station. Address listings of radio amateurs throughout the world can be

## Table 8.5: Summary of some of the major HF operating awards

| Award | Details |
| --- | --- |
| DX Century Club (DXCC)<br>*For further information, contact:*<br>ARRL (www.arrl.org) | Awarded for submitting proof of contact with at least 100 DXCC countries. Endorsements are available for further countries. There are several categories: Mixed (i.e. any mode); Phone; CW; RTTY;; 160 metre; 80 metre; 40 metre; 6 metre; 2 metre; Satellite; Five band DXCC (100 countries on each of five bands. There is an honour roll for those who have contacted within 10 of the maximum number of countries available at any given time. |
| IARU Region 1 Award<br>*Can be obtained via:*<br>Radio Society of Great Britain (www.rsgb.org) | This may be claimed for producing evidence of having made contact with stations in countries that are members of Region 1 Division of the International Amateur Radio Union. There are three classes of the award: Class 1 for contacts with all member countries; Class 2 for 45 member countries; Class 3 for 30 member countries. |
| Islands on the Air (IOTA)<br>*For further information, contact:*<br>Radio Society of Great Britain (www.rsgbiota.org) | This is awarded for making providing proof of contact with stations located on islands (worldwide and regional). The aim of the programme is to encourage contacts with island stations around the world and, encourage activity on these islands since so many people live on them. The basic award is for 100 islands, but higher achievement awards are available for 200, 300, 400, 500, 600 or 700 islands. The latest addition to the range is a 1000 Islands Trophy, also with additional Shields. |
| Worked All Britain<br>*For further information, contact:*<br>Worked All Britain Group (www.wib.intermip.net) | The Worked All Britain Awards Group (W.A.B.) was devised by the late John Morris G3ABG in 1969. The award was devised to promote an interest in Amateur Radio in Britain and to sponsor a series of awards based on the geography of Great Britain and Northern Ireland. This award is available for contacts (it is also available on a "heard" basis for short wave listeners) with UK National Grid 10 km squares which have reference consisting of two letters followed by two numbers, e.g. TQ99. The award is given in different number of different categories, details of which can be found on their website. |
| Worked All Continents<br>*Can be obtained via:*<br>Radio Society of Great Britain (www.rsgb.org) | This is issued by IARU Headquarters for providing evidence of contacts with amateur radio stations in each of the six continents (North America, South America, Europe, Africa, Asia and Oceania). For UK stations cards may be sent to the RSGB HF Awards manager who will supply a certified claim to the IARU Headquarters at ARRL headquarters. |
| Worked All Zones<br>*For further information, contact:*<br>CQ Magazine (www.cq-amateur-radio.com) | This is awarded for confirmed contacts with stations in each of the 40 CQ Zones. A five band version is also available. |
| Worked ITU Zones<br>*Can be obtained via:*<br>Radio Society of Great Britain (www.rsgb.org) | This award is available for contacts with states in 70 of the 75 ITU Zones. |

bought from the RSGB on CD-ROM. Alternatively there are a number of web sites where it is possible to obtain this data (e.g. www.qrz.com), although addresses may not be completely up to date.

A number of stations use QSL managers to handle their cards for them. A station on a remote island or another inaccessible location may not have a reliable or frequent postal service. Accordingly, it is far more convenient to send logs to someone in a country with all the required facilities to act as the manager. DXpeditions also need a person to whom cards can be sent, as the operation will be over once the cards arrive. Details of QSL managers are usually mentioned occasionally during the operation, so it is worth listening out for this information. Additionally, details are published in magazines such as RadCom and on DX web sites.

## Awards

Many amateurs like to gain some of the many operating awards that are available. These present a challenge, and once the awards have been received, they look very attractive and can be mounted in picture frames and displayed in the shack.

There is a wide variety of awards but one of the most famous is the DX Century Club Award. This is awarded by the American Radio Relay League (the US national amateur radio society) to people who can supply proof (QSL cards) that they have made contact with stations in 100 'entities' (this usually means countries but see the rules). For people who have contacted more, there are endorsements that can be added. Some people have made contact with over 300 entities, and this represents a considerable achievement. A summary of some of the major awards is given in Table 8.5.

## Contests

Contests can be an excellent way of contacting rare stations in new countries. During the major contests, the bands come alive with stations, and many people and groups travel to different countries or islands to activate a more-sought-after country or prefix. Furthermore, some stations in rare countries come on the air just for the contest. Given that contest contacts are very fast, so that as many stations as possible can be contacted, it all makes an ideal hunting ground for rare and interesting stations. It is by no means uncommon for people to contact 100 countries or more during a contest weekend.

There are many contests during the year. The main ones include the ARRL DX Contest, CQ WPX, CQ Europe, All Asia DX Contest, IOTA and the CQ World Wide Contest. Of these, possibly the CQ World Wide is the most popular, and there are generally a good number of DXpedition stations active specially over the weekend of this contest. Also, for each of these contests, there is a phone or SSB leg, and a Morse leg on different weekends, often about a month apart.

When operating in a contest, be prepared to make contacts very quickly. Everyone wants to make as many contacts as possible and there is no time for exchanging pleasantries. I find it helpful to have the logbook in a position where entries can be made as quickly as possible and to have some spare paper

## Table 8.6: Major Amateur Radio HF Contests

| Contest | Date | Comments |
| --- | --- | --- |
| CQ-Worked PrefiXes (WPX)(RTTY) | 2nd full W/E Feb | Stations contact as many stations as possible. Extra points given for new prefixes contacted. |
| ARRL DX Contest (CW) | 3rd Full W/E Feb | Stations contact USA/Canada |
| ARRL DX Contest (SSB) | 1st full W/E March | Stations contact USA/Canada |
| CQ-Worked PrefiXes (WPX)(SSB) | Last full W/E March | Stations contact as many stations as possible. Extra points given for new prefixes contacted. |
| CQ-Worked PrefiXes (WPX) (CW) | Last Full W/E May | Stations contact as many other stations as possible. Extra points are given for new prefixes that are contacted. |
| CW Field Day (UK) (CW) | Usually 1st W/E June | British portable stations make as many contacts as possible. |
| All Asia (CW) | 3rd full W/E June | Contact stations in Asia. |
| IARU-HF Championship (SSB / CW) | 2nd full W/E July | Contact as many stations as possible. Extra points given for new countries contacted. |
| Island On The Air (SSB/CW) | Last full W/E July | Contact as many stations as possible. Extra points given for working island stations. |
| Worked All Europe-DX (CW) | 2nd full W/E Aug | Stations outside Europe contact as many European stations as possible. |
| All Asia (SSB) | 1st full W/E Sept | Contact stations in Asia. |
| SSB Field Day (SSB) | 1st full W/E Sept | Portable stations make as many contacts as possible. |
| Worked All Europe-DX (SSB) | 2nd full W/E Sept | Stations outside Europe to contact as many European stations as possible. |
| CQ-WorldWide (RTTY) | 4th full W/E Sept | Contact as many stations in as many countries as possible. |
| CQ-WorldWide (SSB) | Last full W/E Oct | Contact as many stations in as many countries as possible. |
| Worked All Europe-DX (RTTY) | 2nd full W/E Nov | Stations outside Europe to contact as many European stations as possible. |
| CQ-WorldWide (CW) | Last full W/E Nov | Contact as many stations in as many countries as possible. |

handy for jotting down any callsigns. Others prefer to use a computer logging system.

If you have not operated in a contest before, take a while to listen to the contacts being made to find out how it works. Normally, the station who 'owns' the frequency will put out a short CQ call and wait for replies. When he hears a station, he will mention his callsign and give a report and contest serial number. The other station will respond by confirming reception of the report and serial number by saying "roger" or "QSL" and then give a report and serial number back. Finally, the first station will confirm reception and say "73s and QRZ".

In most contests, the idea is to contact as many other stations as possible. However, points may be gained in a number of ways. They may be given for each station contacted, but there may be more if they are in another country or continent, and then multipliers may be given for the number of different countries (or zones etc) contacted. Each contest has its own rules and they differ from one to the next.

Contest report numbers also vary. In some contests, they may be the serial number of the contact, starting from 001, or in others it may be the zone in which the station is located, while in others it may be the power being used. In one contest, the operator's age is sent (ladies send 00!). A summary of the major contests is given in Table 8.6.

It is also worth having a good look around the bands during a contest. Sometimes those that may normally appear closed will suddenly open to support communication. The reasons for this are two-fold. First, many of the 'big' stations with large antennas come on the bands and can be heard. Second, more people stay up around the clock to make contacts whereas at other times of the year they would not be active from that particular region at that time of day or night. It is therefore worth spending some time assessing what activity levels are like.

Part of the skill of operating in a contest is knowing when to stick at contacting a station and when to move on. Competition is usually fiercest at the beginning. However, towards the end, for example on the second day of a two-day contest, the pile-ups may be smaller as many people will have made contact with the rare stations. Accordingly, Sunday afternoon and evening can be very rewarding times on the bands, although some pile-ups can still be large and difficult to get through.

## Further reading:
*Amateur Radio Operating Manual,* 6th edn, ed Don Field G3XTT, RSGB, 2004.
*RSGB Yearbook* (published annually by the RSGB).

## Websites of interest for contesting:
RSGB HF Contests can be found at: www.rsgbhfcc.org
ARRL site at: www.arrl.org/contests

CHAPTER 9

# Setting up the Radio Station

*In this chapter:*

- Requirements for the shack
- Ideas for locations for the shack
- Constructing a table
- Power distribution
- Lighting
- Choosing equipment
- Equipment list
- Connecting the station up
- Equipment layout
- Decorating the shack
- Safety

IT is important to have a radio area ('shack') that is well planned. Since a lot of time is likely to be spent there, listening, transmitting, constructing or doing any one of a number of tasks, it is necessary that the room is easy to use and comfortable. If it is difficult to use, then enjoyment will be reduced considerably, but if it is a well-set-out shack it will bring more enjoyment and encourage more use of the equipment and space.

The first step in setting up a shack is to determine where it will be. This is not always as easy as it might seem. While the ideal location would be a spare room, this is not always feasible. Fortunately, it is possible to set up a shack in many different places around the home by using a little imagination and ingenuity – walk-in cupboards, garden sheds, backs of garages and a whole host of other places can be used.

## Shack requirements

Everyone will have their own ideas about how they want their shack to be set out. Some people will want to use it primarily for operating, while others will want the space for construction, and some will want it to cater for both. This makes hard and fast rules about the ideal shack difficult to define. The only principle that can be applied to every shack is that it pays to put in some thought and planning before actually setting it up.

During these initial stages of planning, there are a number of common points that should be considered. Features to be investigated should include the amount of space required, availability of power and access for feeders.

# HF AMATEUR RADIO

**Photo 9.1: G4BWP keeps all of his equipment within easy reach**

You must also decided whether to shut the shack off from the rest of the house. If radio equipment is located in the living area, it will certainly lead to comments from the rest of the family about the noise. On top of this, it is not wise to leave the equipment open to others, particularly if there are children around.

Although it is useful to have the shack separate from the rest of the house, it should still be easily accessible. This is because it is often nice to spend just a few minutes in there to see if the 15m band is open, for example, or to find out if that rare DX station is around. If the shack is difficult to get to, there is a tendency not to bother to use it and miss out on the wanted DX.

Another factor to consider is the necessity of getting feeders in and out. Although coaxial cables can be run around the house without too much difficulty, they look rather unsightly. In view of this, the shack should have ample access to the outside. This becomes even more important if open-wire feeders are being used.

Size is another important feature. In most cases, the actual operating space is only part of what is required, the rest being taken up with components, surplus equipment, books and all the rest of the paraphernalia which goes with every amateur station. This means that the shack has to be large enough to accommodate not only the equipment itself, but also the rest of the trappings, as well as somewhere for the operator to sit.

## What's available

After deciding on the requirements of the shack, the next stage is to look at what is available. Unfortunately, the various places available seldom live up to what one would like. This is where a little ingenuity comes in useful to make the best out of what is available.

One possibility for the site of a shack is a large cupboard. At first sight, this might not seem to be a particularly good solution because of the space limitations. However, if it is well planned, making the best use of all the space that

# CHAPTER 9: SETTING UP THE RADIO STATION

**Photo 9.2: G4CAO operates from what used to be his 'smallest room'**

is available, it can prove to be perfectly acceptable. A table top can be built in to the cupboard, but remember to leave sufficient space behind the surface to feed up cables without having to remove the plugs. Also remember that access for power and antenna feeders will be required without undue intrusion into the rest of the home.

Another possibility for a shack may be an outside shed. These are not always ideal because they can become very cold in winter, they require power to be installed and security may be a problem, but they do have a number of advantages. Antennas can often be located quite close by, the radio equipment is kept out of the house, and they can provide a good self-contained room for all the equipment. If this option is taken up, then it is well worth lining the shed to ensure it remains warm in the winter.

Some people have used a convenient corner in a garage. In many ways, this is not ideal, but is may be possible to convert an area of the garage, and even partition it off from the rest of the space so that it becomes warmer and more comfortable.

For many people, one of the most logical and convenient places to set up their shack will be in the loft or attic. Having the shack there has several advantages. For example, a loft or attic can often be reasonably easy to convert into a shack, and it will usually be fairly spacious, giving plenty of storage room. Another advantage is that the shack will be separate from the rest of the house, while access to it is reasonably easy.

However, there are a few disadvantages that should be considered before making the final decision. The first and most obvious question to ask is whether or not the loft timbers will stand the weight of all the equipment and people who may go up there. If there is any doubt over this, it is worth consulting a builder or surveyor for his opinion. The cost will also have to be considered – a floor will have to be put down, a loft ladder installed as well as having to install power and any other work that may be needed. Also, remember that a loft will suffer from large variations in temperature – in summer, it will become very hot and in winter it can be very cold.

Unfortunately, there are only a comparatively small number of people who possess a spare room that can be devoted to amateur radio. However, if one can be used, then it can be a very convenient option to choose. Not only will it be warmer and more comfortable, but it will be much easier to spend the odd five minutes in there as it will not necessitate a trip down the garden to the shed, going out into the cold garage or whatever. On top of this, it is likely to require less work in getting the shack set up. There will be no problems of having to fit a floor as in a loft, or line the walls as in a shed. Also there should be sufficient power already installed.

## Making a shack table

Wherever the shack is located, some form of table or desk will be required. One solution is to look in the shops for a sturdy desk. This can either be obtained new, but a much cheaper option is to look in a second-hand or surplus office furniture shop. Computer tables can also make stylish operating desks.

If it is not possible to obtain exactly what is required, then it is not difficult to build a suitable table, even for those with limited woodworking skills. Remember to keep the design simple and functional and takes care during construction.

**Fig 9.1:** Construction of a table

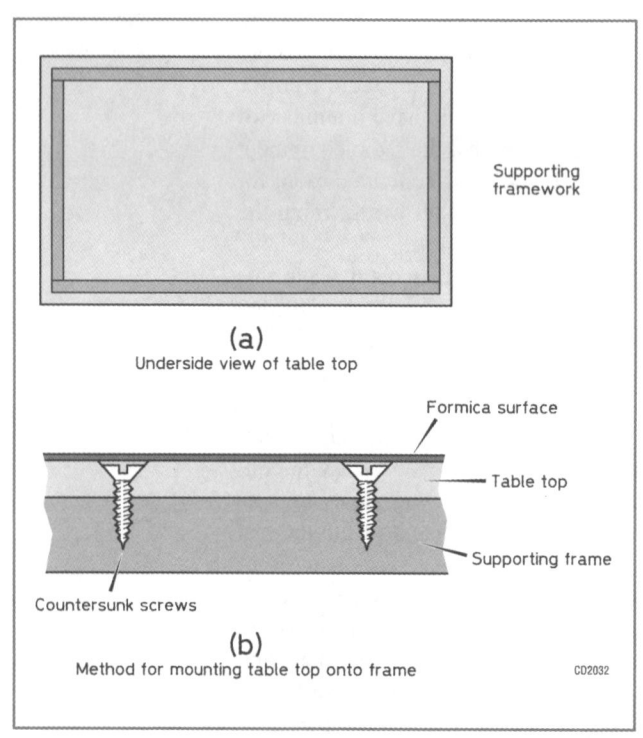

There are a variety of ways to make a table. The main requirements are that it should be large enough and sufficiently sturdy. Even with modern equipment, the weight soon builds up if a few pieces of equipment are added to the station. The first step in the construction is to set the size. The depth should be sufficient to allow space behind the equipment for cables, remembering coaxial cable has a finite bending circle - normally about 3in should be sufficient. Then there should be adequate space in front of the equipment to allow easy operation of the equipment – allow around 15in for this as a guide. This will give space for a log book, microphone, Morse key and room to rest your arm to make for easy operating during long contests.

The work surface can be made from a sheet of blockboard or plywood, although chipboard or MDF can be used if is of sufficient thickness. The table top can be strengthened by a framework of 2in by 1in wood to give extra support and prevent any tendency to sag. This can sometimes be a problem, especially if heavy equipment remains on the table for a long period of time. This framework can be fixed to the table top using countersunk screws as shown in Fig 9.1. Although this does leave holes in the work top they can be filled and then the whole surface can be covered with Formica.

The framework enables the legs to be attached more easily, in addition to providing extra support for the table top. There are a variety of ways in which this can be done. Wooden legs can be attached to the framework, but remember to have "struts" as shown to provide support at the bottom of the legs. There are many other ingenious ways of making suitable legs for the table. For example, it may be possible to use the steel legs from an office desk that is beyond repair.

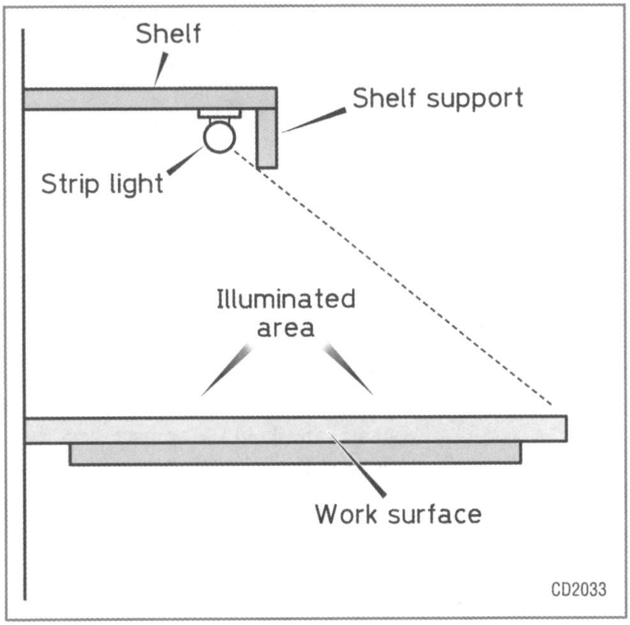

Fig 9.2: Lighting under a shelf can be used to illuminate the work area. Be careful to follow the manufacturer's instructions when fitting and do not mount it too close to the shelf support or the equipment.

## Lighting

Lighting is an important feature in any shack, especially if any construction work is envisaged. One way of improving the lighting is to install a small strip lamp under a shelf above the table top to illuminate the work area. The front support on the shelf can then be used to shade the lamp from direct view. When deciding exactly where to fix it be careful to ensure that the manufacturer's recommendations for fitting and ventilation are obeyed.

An anglepoise-type lamp can be used if further lighting is required. These are ideal as they enable a large amount of light to be concentrated on the required area.

## Power distribution

Some thought should be given to the way the mains power is distributed to the various pieces of equipment in the shack. The number of items requiring mains power can be surprisingly high and, if a single socket is used, this will quickly become overloaded. It is therefore advisable to use one or more of the mains distribution blocks containing four or five outlets in a straight line. These can be fastened to a suitable place near the back of the table, either on top or below the work surface. Then each piece of equipment can be permanently plugged in and turned on from its front-panel switch as required.

The cables supplying these distribution blocks can be taken back to a common switch or circuit breaker so that the whole station can be turned off quickly and easily. It is also very wise to include a residual current circuit breaker. These switch off the power when there is an imbalance in the levels of current taken between the live and neutral power lines, providing a very useful safety feature. They can be bought quite cheaply from most hardware or electrical suppliers.

A further item to consider is a single main switch to turn the whole station off. When the station is to be closed down, this can be used to isolate all density.

## Choosing equipment

There is an enormous choice of equipment available on the market. From small hand-held FM transceivers right up to the multi-mode, multi-band transceivers, there is something for almost every requirement. Choosing the best equipment for a particular station is not always easy. It is best to look carefully through the magazines to see what is available from the dealers or second-hand in the readers' advertisements. Also, read the reviews to find out about the individual pieces of equipment.

It is also worth considering what is needed for the shack – the modes of operation envisaged, the power levels required, frequency coverage needed etc. In this way, the field can be narrowed down.

It is sensible to look at a variety of equipment before parting with any hard-earned cash. Mobile rallies, hamfests and the like are a very good opportunity for this. Many dealers come to these events and it is possible to look at a good number of units to see what they are like in real life rather than in pictures in magazines. It may also be worth a visit to a dealer – here it is often possible to sit down with a set, use it for a while and see how it 'feels' and whether it is the right one for you.

You should take extra care when buying second-hand equipment. It may have been in use for some time and might need attention – this is particularly true when buying privately. Most dealers have a name to protect and will not offer doubtful equipment for sale but this may not be the case with a private transaction. Check over the general appearance of the set – has it been well cared for or heavily used? First impressions count for a lot. If it has been heavily used, some components like switches may be worn and in need of replacement. Components may also have been replaced inside, and there may always be the possibility of further problems if a repair has not been completed properly. Also check to see if any modifications have been undertaken. If so, have they been implemented well? Often modifications are done by people who do not have sufficient skill with a soldering iron, and they may cause more damage to the set than the modification is worth.

It is also necessary to check the general performance of the set. Does it appear to be sensitive – can many stations be heard and at good strength? Are any beacons audible at about the right strength? Do the switches work properly, or are they intermittent if they are touched? Is the tuning smooth? Is the

# CHAPTER 9: SETTING UP THE RADIO STATION

power output correct, and if a contact is made using a transmitter or transceiver, do the other stations report a good-quality signal?

It is often worth taking a friend along with you, especially if it is a private sale. A second opinion is always valuable and, if they have been in the hobby for some time and have experience of buying equipment, it can be helpful to draw on this.

While it is necessary to be careful about buying equipment, most people are honest and there are many good deals to be found.

## Equipment list

A variety of items will be needed for the shack. Obviously, the precise equipment needed will depend on your plans for the station. However, the shopping list below is a good starting point for a basic station:

- HF transceiver (or separate transmitter and receiver)
- Power supply for the transceiver (if one is not supplied). Make sure it can supply sufficient current for the transceiver.
- Ferrite rings for interference suppression (see below)
- Antenna tuning unit
- VSWR meter (if one is not already included in the transceiver)
- Patch cables to connect the transceiver to the VSWR meter and then the ATU (remember to buy the right connectors for each connection)
- Microphone (remember that the correct connector will be needed for the particular transceiver being used as the type of connector or wiring can vary from one manufacturer to another)
- Morse key (again remember the connector)
- Loudspeaker and / or headphones (again check on the connector required for the loudspeaker). Note that often a loudspeaker is contained within the transceiver.
- Antenna (don't forget the connector to connect the feeder to the ATU)
- Logbook, scrap paper for notes and a pen
- Clock
- Prefix list
- World map

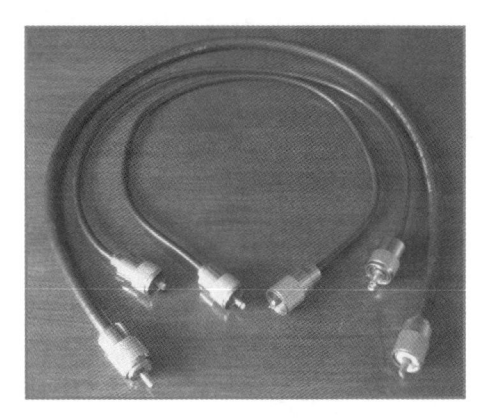

Photo 9.3: PL259 patch leads suitable for connecting between the transceiver and VSWR meter and VSWR meter and ATU.

## Connecting the station up

There are many ways in which the equipment in an amateur radio station can be connected up. To an extent it will depend on the equipment being used, but there are some basic guidelines that can be followed whatever the set-up. Unfortunately, the electrical connections may not always fit in with the most ergonomic use of the equipment, but wires are flexible and can often be made to length.

A basic diagram for the way in which a basic station must be connected is shown in Fig 9.3. Basically, the power supply for the transceiver is connected

# HF AMATEUR RADIO

**Fig 9.3: Connecting equipment in an HF amateur radio station**

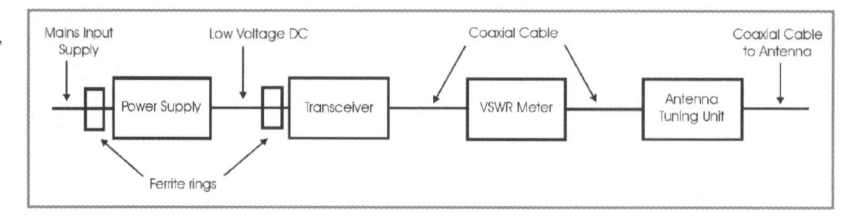

to the power source, typically the mains. The power supply is connected in to the transceiver which in turn is connected to a VSWR meter. This measures the actual VSWR being seen by the transceiver. This is very important because transistor output stages do not like high levels of VSWR. While is useful to reduce the VSWR between the ATU and the antenna, the crucial thing is to reduce the VSWR seen by the transceiver output.

The diagram shows ferrite rings being used around the leads to and from the power supply. These reduce the level of RF leaking into the mains supply. RF leakage can cause interference to other items of mains powered electronics equipment such as hi-fi sets and televisions. If power supply leads have moulded on plugs, then it is possible to use a clip-on ferrite as an acceptable alternative. This does not however remove the need for a good RF earth such as a copper pipe driven well into the ground.

## Equipment layout

While ideas for the design of the shack and table are being formulated, it is well worth giving some thought to the layout of the equipment itself so that the whole station comes together properly. The main point is that the equipment can easily be reached with the minimum of arm ache.

It is important that the tuning knob on the transceiver is correctly placed. This should be a couple of inches above the table surface and in a position

**Photo 9.4: Experimenter G4JNT has his test equipment neatly rack mounted. The use of a swivel chair ensures that everything is to hand**

Photo 9.5: G0MYX stacks his gear three shelves high. His work bench can just be seen on the right at 90° to the operating position

where it can be reached easily. The microphone and Morse key should also be placed where they are easy to use. Often, a microphone will be held with the left hand (for those who are right handed), leaving the right hand free to write notes. Similarly the Morse key, if one is used, will be operated using the right hand, and so it should be placed on the right-hand side of the table. It is also convenient to have space for a note pad and the logbook. Computers are often used in stations today for a variety of purposes from logging to connecting to the DX Packet Cluster or for propagation prediction software. Consideration should be given to locating the keyboard and screen so that they can also be used easily.

A typical station layout may have the main transceiver in the middle at the front with a linear amplifier to one side. A second receiver may be set to the other side. Ancillary equipment such as the ATU and VSWR bridge can be placed on a shelf over the transceiver and other main units. In this way, they are easy to adjust and an eye can be kept on the meter readings.

However, no two people's needs are the same and so each station will need to be planned on its own merits. To gain a few extra ideas, it is worth looking at photographs of other stations that appear in the magazines quite regularly. Stations of the top DXers are particularly useful because they will have spent many hours in their shacks using the equipment. On the way, they are likely to have discovered many of the pitfalls of poor layout and reorganised them to be very easy and comfortable to operate. Finally, it is well worth investing in a good chair – it is not easy to relax in the shack, or spend several hours operating in a contest, if the chair is uncomfortable.

## Decorating the shack

Once the equipment in the shack has been set up, the walls can be decorated. Maps like great circle maps, QRA locator maps, or prefix maps are very use-

# HF AMATEUR RADIO

**Photo 9.6: The neatly laid out station of GU4YOX enables him to use the Morse key and computer keyboard simultaneously**

ful and give a lot of visual information very quickly. These should be easy to see, and may be on the wall behind the equipment.

It is also nice to put up some QSL cards, particularly those showing the best DX that has been worked. Unfortunately, mounting them on the wall can sometimes damage them. Pinning them to the wall obviously puts a hole in them, and Blutack can leave a mark after a while.

To overcome this problem, it is possible first to mount the cards onto a postcard or some other suitable card using photograph corners. Then the postcard can be pinned or stuck to the wall, leaving the QSL cards free from damage. Alternatively, transparent 'wallets' are obtainable from some photography suppliers.

## General safety

It is probably true to say that safety standards in shacks have improved over the past few years. This is partly as a result of an increased awareness of the hazards, and partly due to the voltages in equipment being lower. Another reason is that more commercially made equipment is being bought and this has to comply with safety standards. However, there are still a number of safety precautions which can be easily incorporated into the shack.

You should ensure that all equipment is properly earthed. Sometimes there is a tendency to leave the earth connection off some pieces of equipment, which can be dangerous because it means that the whole of the case can rise to mains potential under certain fault conditions.

Earth leakage or residual current circuit breakers have already been mentioned, along with a single off switch for the whole station. These are well worth fitting.

While on the subject of electrical safety it is worth pointing out that all equipment carrying hazardous voltages or high levels of RF should be

# CHAPTER 9: SETTING UP THE RADIO STATION

**Photo 9.7:** Newport Amateur Radio Society's station can be folded away into a cupboard when not in use

enclosed in cabinets. This is particularly important if visitors are likely to come into the shack at any time.

Finally, radiated RF should be kept away from inhabited areas of the house. Although it is unlikely that low power and non-directional antennas could cause any harm, this may not be true where high power and directional antennas are used. However, as it is difficult to assess field strengths at a particular place it is best to keep all RF at a distance.

These ideas represent only a few of the ways for keeping a shack safe. There are many other things you can do to make it safer. But the best advice is to have a general awareness of the dangers that might arise. Then the shack will be a safer place for you and any visitors who may call in.

## Further reading
*Amateur Radio Operating Manual*, 6th edn, ed Don Field G3XTT, RSGB, 2004.

# HF AMATEUR RADIO

# APPENDIX

# Abbreviations and Codes

*In the appendix:*

- Commonly used abbreviations
- Q code
- Phonetic Alphabet
- RST code
- Morse code

### Table A.1: Commonly used abbreviations

| | |
|---|---|
| ABT | about |
| AGN | again |
| AM | amplitude modulation |
| ANT | antenna |
| BCI | broadcast interference |
| BCNU | be seeing you |
| BFO | beat frequency oscillator |
| BK | break |
| B4 | before |
| CFM | confirm |
| CLD | called |
| CIO | carrier insertion oscillator |
| CONDX | condition |
| CPI | copy |
| CQ | a general call |
| CU | see you |
| CUAGN | see you again |
| CUD | could |
| CW | carrier wave (often used to indicate a Morse signal) |
| DE | from |
| DX | long distance |
| ERE | here |
| ES | and |
| FB | fine business |
| FER | for |
| FM | frequency modulation |
| FONE | telephony |
| GA | good afternoon |
| GB | goodbye |
| GD | good |
| GE | good evening |

### Table A.1: Commonly used abbreviations *(continued)*

| | |
|---|---|
| GM | good morning |
| GN | goodnight |
| GND | ground |
| HBREW | home brew |
| HI | laughter |
| HPE | hope |
| HR | here |
| HV | have |
| HW | how |
| LID | poor operator |
| LW | long wire |
| MOD | modulation |
| ND | nothing doing |
| NW | now |
| OB | old boy |
| OM | old man |
| OP | operator |
| OT | old timer |
| PA | power amplifier |
| PSE | please |
| R | roger (OK) |
| RCVD | received |
| RX | receiver |
| RTTY | radio teletype |
| SA | say |
| SED | said |
| SIGS | signals |
| SRI | sorry |
| SSB | single sideband |
| STN | station |
| SWL | short wave listener |
| TKS | thanks |
| TNX | thanks |
| TU | thank you |
| TVI | television interference |
| TX | transmitter |
| U | you |
| UR | your, you are |
| VY | very |
| WID | with |
| WKD | worked |
| WUD | would |
| WX | weather |
| XMTR | transmitter |
| XTAL | crystal |
| XYL | wife |
| Z | GMT – the letter is added after the figures i.e. 1600Z is 16 00 hrs GMT |
| YL | young lady |
| 73 | best regards |
| 88 | love and kisses |

# APPENDIX: ABBREVIATIONS AND CODES

## Table A.2: Q Code

| | | | |
|---|---|---|---|
| QRA | What is the name of your station? The name of my station is | QRT | Shall I stop sending? Stop sending |
| QRG | What is my frequency? Your exact frequency is .......... | QRU | Do you have any messages for me? I have nothing for you |
| QRL | Are you busy? I am busy | QRV | Are you ready to receive? I am ready |
| QRM | Is there any (man made) interference? There is (man made) interference | QRZ | Who is calling me? You are being called by ........ |
| QRN | Is there any atmospheric noise? There is atmospheric noise | QSL | Can you acknowledge receipt? I acknowledge receipt |
| QRO | Shall I increase my power? Increase power | QSP | Can you relay a message? I can relay a message |
| QRP | Shall I reduce power? Reduce power | QSY | Shall I change to another frequency? Change to another frequency |
| QRQ | Shall I send faster? Send faster | QTH | What is your location? My location is |
| QRS | Shall I send more slowly? Send more slowly | QTR | What is the exact time? The exact time is |

## Table A.3: Phonetic Alphabet

| | | | | | | | |
|---|---|---|---|---|---|---|---|
| A | Alpha | H | Hotel | O | Oscar | V | Victor |
| B | Bravo | I | India | P | Papa | W | Whisky |
| C | Charlie | J | Juliet | Q | Quebec | X | X-ray |
| D | Delta | K | Kilo | R | Romeo | Y | Yankee |
| E | Echo | L | Lima | S | Sierra | Z | Zulu |
| F | Foxtrot | M | Mike | T | Tango | | |
| G | Golf | N | November | U | Uniform | | |

## Table A.4: RST Code

**Readability**
1 Unreadable
2 Barely readable
3 Readable with difficulty
4 Readable with little difficulty
5 Totally readable

**Strength**
1 Faint, barely perceptible
2 Very weak
3 Weak
4 Fair
5 Fairly good
6 Good
7 Moderately strong
8 Strong
9 Very strong

**Tone**
1 Extremely rough note
2 Very rough note
3 Rough note
4 Rather rough note
5 Modulated note
6 Near d.c. note but with smooth ripple
7 (d.c. note of ripple)
8 Good d.c. note with a trace of ripple
9 Pure d.c. note

## Table A.5: Morse Code

| | | | |
|---|---|---|---|
| A | ._ | N | _. |
| B | _... | O | ___ |
| C | _._. | P | .__. |
| D | _.. | Q | __._ |
| E | . | R | ._. |
| F | .._. | S | ... |
| G | __. | T | _ |
| H | .... | U | .._ |
| I | .. | V | ..._ |
| J | .___ | W | .__ |
| K | _._ | X | _.._ |
| L | ._.. | Y | _.__ |
| M | __ | Z | __.. |
| 1 | .____ | 6 | _.... |
| 2 | ..___ | 7 | __... |
| 3 | ...__ | 8 | ___.. |
| 4 | ...._ | 9 | ____. |
| 5 | ..... | 0 | _____ |

### Punctuation

| | |
|---|---|
| Full Stop (.) | ._._._ |
| Comma (,) | __..__ |
| Question Mark (?) | ..__.. |
| Equals sign (=) | _..._ |
| Stroke (/) | _.._. |
| Mistake | ........ |

### Procedural Characters

[For procedural characters made up of two letters they are sent as a single letter with no break between them.]

| | |
|---|---|
| Start of Work (CT) | _._._ |
| Invitation to Transmit (KN) | _.__. |
| End of Work (VA) | ..._._ |
| End of Message (AR) | ._._. |
| Invitation to Transmit (K) | _._ |
| Invitation to a Particular Station to Transmit (KN) | _.__. |

### Spacing and length of elements

A dash is equal to three dots.
The space between elements which form the same letter is equal to one dot.
The space between two letters is equal to three dots.
The space between two words is equal to seven dots.

# Index

## A
A index . . . . . . . . . . . .22
Abbreviations, commonly
   used . . . . . . . . . . . . .133
Alphabet, phonetic . . . .135
Amateur bands . . . . . . . .98
Amphenol coaxial connector
   . . . . *see UHF connector*
Amplitude modulation
   (AM) . . . . . . . . . . .25,37
AMTOR . . . . . . . . . . .4,30
Angle of radiation . . . . .13
Antenna tuning unit
   (ATU) . . . . . . . . . . . . .79
Antennas . . . . . . . . . . .2,73
   Location . . . . . . . . . . .95
Ap index . . . . . . . . . . . .22
Atmosphere, structure of .8
Audio filters . . . . . . . . .42
Audio frequency shift key-
   ing (AFSK) . . . . . . . .28
Audio image . . . . . . . . .37
Automatic gain control
   (AGC) . . . . . . . . . . . .43
Automatic level control
   (ALC) . . . . . . . . . . . . .66
Automatic request for
   repeat (ARQ) . . . . . .31
Awards . . . . . . . . . . . .118

## B
Balun . . . . . . . . . . . . . .84
Band plans . . . . . . . . . .97
Bands, amateur . . . . . . .97
   UK HF band list . . . . .4
   10m band . . . . . . . . .105
   12m band . . . . . . . . .104
   15m band . . . . . . . . .104
   17m band . . . . . . . . .103
   20m band . . . . . . . . .101
   30m band . . . . . . . . .101
   40m band . . . . . . . . .100
   80m band . . . . . . . . . .99
   160m band . . . . . . . . .98
Baudot code . . . . . . .29,30
Beam antennas . . . . .80,93
Beat frequency oscillator
   (BFO) . . . . . . . . . . .25,39
Binary phase shift keying
   (BPSK) . . . . . . . . . . . .32
Blocking, receiver . . . . .46
BNC connector . . . . . . .78
Break-in . . . . . . . . . . . .58

## C
Capture effect . . . . . . . .28
Carrier . . . . . . . . . . . . .33
Characteristic impedance . .
   . . . . . . . . . . . . . . . . .76
Chordal hop . . . . . . . . .17
Clipping, speech . . . . . .60
Coaxial feeder . . . . . . . .74
Codes
   Baudot . . . . . . . . . .29,30
   Morse . . . . . . . . . .23,136
   Q . . . . . . . . . . . . . . .135
   RST . . . . . . . . . . . . .135
Compression, speech . . .60
Connectors . . . . . . . . . .76
   BNC . . . . . . . . . . . . .78
   N-type . . . . . . . . . . . .78
   UHF . . . . . . . . . . . . .77
Contacts, basic . . . . . .107
Contests . . . . . . . . . .1,118
Counterpoises . . . . . . . .95
CQ call . . . . . . . . . . . .108
Critical frequency . . . . .15
Cross-modulation . . . . .46
Crystal filters . . . . . . . .42
CW . . . . . *see Morse code*

## D
D layer . . . . . . . . . . . . . .9
dBd unit . . . . . . . . . . . .80
dBi unit . . . . . . . . . . . .80
dBW unit . . . . . . . . . . .63
Dead zone . . . . . . . . . . .14
Decorating the shack . .129
Deviation, FM
   transmission . . . . . . .27
Digital signal processing
   (DSP) . . . . . . . . . . . .49
Dipole antenna . . . . . . .82
   Building a dipole . . . .83
Direct-conversion
   receiver . . . . . . . . . : . .36
Direct Digital Synthesizers
   (DDS) . . . . . . . . . .39,41
Directivity of antennas . .80
Disturbance, solar . . . . .18
Doublet antenna . . . . . .89
DX Century Club
   Award . . . . . . . . . . .118
DX nets . . . . . . . . . . .112
DX Packet Cluster . . . .112
DXing . . . . . . . . . . . .1,111
DXpeditions . . . . . . .2,111
Dynamic range . . . . . . .47

## E
E layer . . . . . . . . . . . . .10
Earth system, antenna . .94
Emergency
   communications . . . . .4
End effect in antennas . .83
End-fed wire antenna . . .81
End-fire antenna . . . . . .82
Equipment
   Choosing . . . . . . . . .126
   Layout . . . . . . . . . . .128
   List of . . . . . . . . . . .127

## F
F layer . . . . . . . . . . . . .10
Fading . . . . . . . . . . . . .15
Feeders . . . . . . . . . .73,122
Fifth-order product . . . .45
Filter, crystal . . . . . . . .42
Fixing connectors to
   cables . . . . . . . . . . . .78
Flux, solar . . . . . . . . . .21
Frequency modulation . .27
Frequency shift keying
   (FSK) . . . . . . . . . . . .28
Frequency synthesizers .39
Frequency tailoring,
   speech . . . . . . . . . . .61
Front-to-back ratio,
   antenna . . . . . . . . . . .81

137

# HF AMATEUR RADIO

## G

G5RV antenna . . . . . . . .90
Gain, antenna . . . . . . .80
Geomagnetic storms . . .19
Grey line propogation . .19
Ground system,
  antenna . . . . . . . . . . .94
Ground waves . . . . . . . . .7

## H

Harmonics, transmitter .64

## I

Iambic-mode keyer . . . .59
Image response . . . . . . .43
Incremental receiver tuning
  (IRT) . . . . . . . . . . . . .65
Indirect synthesizers . . .39
Intermediate frequency
  (IF) . . . . . . . . . . . . . .37
Intermodulation
  receiver . . . . . . . . . . .46
  transmitter . . . . . . . . .64
Internet . . . . . . . . . . . .111
Inverted-V antenna . . . .86
Ionisation . . . . . . . . . . . .8
Ionosphere . . . . . . . . . . .8
Ionospheric sounding . . .14
Isotropic source . . . . . . .80

## K

K index . . . . . . . . . . . .22
Key, morse . . . . . . . . . .24
Keying, transmitters . . .59

## L

Lighting, shack . . . . . .125
Linear amplifiers . . . . . .62
Location, antenna . . . . .95
Logging programs . . . .114
Lower sideband (LSB) . .27
Lowest usable frequency
  (LUF) . . . . . . . . . . . .15

## M

Mains power in shack .125
Maximum usable frequency
  (MUF) . . . . . . . . . . . .15
Mesosphere . . . . . . . . . .8
Minimum discernible signal
  (MDS) . . . . . . . . . . . .47
Mixing . . . . . . . . . . . . .36
Mode A . . . . . . . . . . . .31
Mode B . . . . . . . . . . . .31
Modem . . . . . . . . . . . . .28

Modulation . . . . . . . . . .23
Morse code (CW) . .23,136
Multiband dipole . . . . . .87
Multiple reflections . . . .16
Murray code . . . . . . . . .29

## N

N-type connector . . . . . .78
Narrow-band frequency
  modulation (NBFM) . .27
Noise performance,
  receivers . . . . . . . . . .43

## O

Open-wire feeder . . . . . .75
Operating techniques . .107
Operating table for
  station . . . . . . . . . . .124

## P

Pass band . . . . . . . . . .42
Path losses . . . . . . . . . .15
Peak envelope power (PEP)
  . . . . . . . . . . . . . . . . . .63
Phase-locked loop (PLL)
  synthesizers . . . . . . . .40
Phase noise . . . . . . . . .47
Phonetic alphabet . . . .135
Piezo-electric effect . . . .42
Pile-ups . . . . . . . . . . .113
PL259 connector . . . . . .77
Polar cap absorption . . .18
Polar diagram . . . . . . . .80
Pole, filter . . . . . . . . . .42
Power
  shack mains . . . . . . .125
  transmitter . . . . . . . . .63
Predicting,
  propagation . . . . .21,111
Pressel switch . . . . . . . .58
Product detector . . . . . .26
Propagation, radio wave .7
PSK31 . . . . . . . . . . . .4,32

## Q

Q code . . . . . . . . . . . .135
QRP operation . . . . . . .108
QRP transceiver
  Calibration . . . . . . . . .69
  Construction . . . . . . . .67
  Design . . . . . . . . . . . .66
  In operation . . . . . . . .70
  Parts list . . . . . . . .70,71
QRP transmitters . . . . . .55
QSL cards . . . . . . . .2,114
Quadrature phase shift
  keying (QPSK) . . . . . .32

## R

Radials . . . . . . . . . . . . .95
Radiation storms, solar .19
Radio blackouts . . . . . .19
Radio teletype
  (RTTY) . . . . . . . . . .3,29
Radio wave propagation .7
Receivers . . . . . . . . . . .37
Reciprocal mixing . . . . .47
Reflections, radio wave .10
Repeaters . . . . . . . .28,106
Ribbon feeder . . . . . . . .
  . . . . . . .see Twin feeder
Rotator, antenna . . . . . .81
RST code . . . . . . . . . .135
'Rubber stamp'
  contacts . . . . . . . . . .107

## S

Safety
  Antennas . . . . . . . . . .96
  Shack . . . . . . . . . . .130
Selectivity . . . . . . . . . .41
Sensitivity . . . . . . . . . .43
Shack, radio . . . . . . . .121
  Lighting . . . . . . . . . .125
  Power, mains . . . . . .125
  Table . . . . . . . . . . .124
Shape factor . . . . . . . .42
Short-wave bands . . . . .4
Sidebands . . . . . . . . . .26
Single sideband (SSB) . .26
  Transmitters . . . . . . . .57
Skip distance . . . . . . . .14
Skip zone . .see Dead zone
Sky waves . . . . . . . . . . .8
Sloper antenna . . . . . . .85
Slow-scan television
  (SSTV) . . . . . . . . . . . .32
SO239 connector . . . . . .77
Solar disturbance . . . . .18
Solar flux . . . . . . . . . . .22
Solar radiation storms . .19
Speech processing . . . .59
  Clipping . . . . . . . . . .60
  Compression . . . . . . .60
  Frequency tailoring . .61
Split frequencies . . . . .113
Spurious signals . . .63,64
Standing wave ratio
  (SWR) . . . . . . . . . . . .79
Station, radio . . . . . . .121
Station, setup . . . . . . .121
Stop band . . . . . . . . . .42
Stratosphere . . . . . . . . .8
Strong-signal handling
  performance . . . . . . .45

Sudden ionospheric
  disturbance (SID) . . . .18
Sunspots . . . . . . . . . . .12
  Sunspot cycle . . . . .12,22
Superhet receiver . . . . .36

## T

Table, shack . . . . . . . .124
Terminal node controller
  (TNC) . . . . . . . . . . . .29
Thermosphere . . . . . . . .8
Third-order intercept . . .45
Third-order
  intermodulation . . . . .45
Third-order products . . .45
Top band . . . . . . . . . . .98
Transceivers . . . . . . . . .62
Transmission lines . . . . .
  . . . . . . . . . .see Feeders
Transmitter power . . . .63
Transmitters . . . . . . . . .55
Trap dipole antenna . . . .88
Trap vertical antenna . . .90
Troposphere . . . . . . . . . .8
Tuned feeder antenna . . . .
  . . . .see Doublet antenna
Twin feeder . . . . . . . . .75

## U

UHF connector . . . . . . .77
Upper sideband (USB) . .27

## V

Varicode . . . . . . . . . . .32
Variable-frequency
  oscillator (VFO) . . . . .55
  Separate VFOs . . . . .65
Velocity factor . . . . . . .76
Vertical antenna . . . . . .92
VOGAD (voice-operated
  gain-adjusting device) .60
VOX . . . . . . . . . . . . . .58
VSWR tolerance,
  transmitter . . . . . . . . .65

## W

Wide-band frequency
  modulation (WBFM) . .27

## Y

Yagi antenna . . . . . . . .93
Yearling receiver . . . . . .50
  Circuit design . . . . . .50
  Construction . . . . . . .51
  Parts list . . . . . . . .54,55
  Setup . . . . . . . . . . . .52

138